ANALOG

The MIT Press Essential Knowledge Series

A complete list of books in this series can be found online at
https://mitpress.mit.edu/books/series/mit-press-essential-knowledge-series.

ANALOG

ROBERT HASSAN

The MIT Press | Cambridge, Massachusetts | London, England

The MIT Press would like to thank the anonymous peer reviewers who provided comments on drafts of this book. The generous work of academic experts is essential for establishing the authority and quality of our publications. We acknowledge with gratitude the contributions of these otherwise uncredited readers.

This book was set in Chaparral Pro by New Best-set Typesetters Ltd. Printed and bound in the United States of America.

Library of Congress Cataloging-in-Publication Data

Names: Hassan, Robert, 1959- author.
Title: Analog / Robert Hassan.
Description: Cambridge, Massachusetts : The MIT Press, [2022] | Series: The MIT Press essential knowledge series | Includes bibliographical references and index.
Identifiers: LCCN 2022000731 (print) | LCCN 2022000732 (ebook) | ISBN 9780262544498 (paperback) | ISBN 9780262371810 (pdf) | ISBN 9780262371827 (ebook)
Subjects: LCSH: Technology—Psychological aspects. | Analog electronic systems—History. | Analog computers—Psychological aspects. | Digital electronics—Social aspects.
Classification: LCC T14 .H29 2022 (print) | LCC T14 (ebook) | DDC 601—dc23/eng/20220328
LC record available at https://lccn.loc.gov/2022000731
LC ebook record available at https://lccn.loc.gov/2022000732

10 9 8 7 6 5 4 3 2 1

For Kate Daw
(1965–2020)

She bought lemon-yellow carnations perfumed with the taste of hard candy, and garden roses purple as raspberry puddings, and every kind of white flower the florist knew how to grow.

—Zelda Fitzgerald, *Save Me the Waltz* (1932)

CONTENTS

SERIES FOREWORD

The MIT Press Essential Knowledge series offers accessible, concise, beautifully produced pocket-size books on topics of current interest. Written by leading thinkers, the books in this series deliver expert overviews of subjects that range from the cultural and the historical to the scientific and the technical.

In today's era of instant information gratification, we have ready access to opinions, rationalizations, and superficial descriptions. Much harder to come by is the foundational knowledge that informs a principled understanding of the world. Essential Knowledge books fill that need. Synthesizing specialized subject matter for nonspecialists and engaging critical topics through fundamentals, each of these compact volumes offers readers a point of access to complex ideas.

ACKNOWLEDGMENTS

Here, the customs of acknowledgment and dedication attach themselves to memory of a time when I was writing a part of this book. Like many writers, I suppose, sometimes I can recall exactly what I was doing when writing a piece of text. In this case it's not so much that the words themselves have any deep or meaningful resonance, but the fact that when I was actually writing them, I experienced a heightened awareness of my context that would stay with me. The words I was playing around with on the screen then were:

> Like a secret code written in invisible ink all over everyday language and life we metaphorically *paint a picture* to describe something to someone, just as we *draw a line in the sand* to indicate the limits of this or that situation. And when speaking about the physical person, when we say that someone is *blossoming* or is *wilting*, we know exactly what is being communicated. Metaphors acting as linguistic objects allow us to focus on a concept in an evolutionary fashion that has worked its way so deeply into our unconscious mind that we need little or no training as to its practical operation.

Now it really does not do to quote oneself—especially within the confines of the same book—but as I say, the words in themselves are not important here; it's when and where they were written, and it's that time and place and one person that I want to acknowledge. Melbourne, where I live, holds the unenviable world record of the longest number of days under the 2020–2021 COVID pandemic lockdowns. One late summer's day in 2020, Kate, my wife, Theo and Camille, our kids, and I were at home working and schooling like millions of others. Kate was then at late-stage cancer, but you wouldn't have known. She said she felt "ninety-five percent" and looked like she always did, beautiful. Kate had been a painter since she was fourteen and it never occurred to her to do anything else. Of course, her many talents led her to eventually becoming not just a much respected and admired artist (Elton John once purchased six of her paintings on the spot during an unannounced visit to a gallery where she was showing) but also a professor and head of the School of Art at the prestigious Victorian College of Arts in Melbourne.

Anyway, when my words were taking shape, I was in a spare room writing and Kate was in her next-door studio painting. The kids were quiet somewhere downstairs and the Cocteau Twins' song "A Kissed Out Red Floatboat" was billowing out of the space between the rooms from a laptop sitting on a table under a window from where the late sun shone red-orange light into the stairhead connection.

I got up from my desk, went downstairs, and brought up two cups of Earl Grey tea. Long ago Kate had said to me: "Never ask if I want tea, just bring it." So, I did. I went into her studio, and she was sitting at her easel, focused on a detail of a flower. I put the cup and saucer on a rickety little three-legged table she had bought as a student on a trip to Venice. Bending down to pick up the previous cup and saucer from the floor, I kissed her on the crown of her head as she got in close into the canvas, her nose almost parallel to a thin brush dipped with lemon-yellow oil paint. A minute or two later I heard a semi-apologetic call of "Thanks!" from the studio when I was back at my chair, as Kate broke off from her concentration to drink the tea.

It's common enough in books to thank loved ones for what they give to you, and that's good. Here I wanted to go just a little further and express an infinite thanks to Kate for this one memory of her that was representative in so many ways of what we had. And what we had provided me with everything, including her advice and support to take the path that took me to writing this book, to everything I wrote before it, and to everything I will write.

1

INTRODUCTION

"Pause! We Can Go Back!"

In his review of David Sax's 2016 book, *Revenge of the Analog: Real Things and Why They Matter*, the environmentalist Bill McKibben wrote: "Digital life's too self-absorbed—either we evolve quickly away from the social primates we have always been or else we will quietly suffer from the solipsism inherent in staring at ourselves reflected in a screen. It's too jumpy: concentration, from which all that is worthwhile emerges, is the great loss."[1]

McKibben is mostly generous in his evaluation. However, he perceives a certain unworldliness in Sax's attitude. He senses that for Sax analog is primarily a neutral technology (as is digital) but that digital somehow lacks the "realness" of analog. And so, the resurgence of certain types of popular analog technologies such as hi-fi systems, vinyl records, board games, and other technologies reflects a peculiar, and almost unaccountable, psychological nostalgia for a predigital time. It's not that simple, McKibben argues, and our relationship to analog and digital technologies can't be distilled down to such a simplistic essence. McKibben doesn't develop his point that there is something much deeper going on, and in fairness he can't be expected to tell a more sophisticated story in the space of a couple of hundred words. His idea is certainly worth developing, though.

By way of an introduction, I want to focus on the quote and develop it a little, so to set the scene for this book. McKibben's two sentences encompass two impressions relating to analog technology that can be used to explore the concept more fully: *evolution* and *loss*. I cite these in McKibben's order, but they have equal billing in my story. *Evolution* anyway seems like a good place to begin. To think of technology as neutral is an unexceptional (if still sometimes controversial) idea in the heads of many people today. It's an idea with its own clearly relatable common sense. It's contained in the refrain, sadly too often heard, that "Guns don't kill people; people do." To hold this view is at a deeper level to believe that humans and technology are separate spheres of existence. Aristotle thought so. Democritus did as well. As did René Descartes, although he also suggested that the body and the brain could be immeasurably improved by technology.

There has been a good deal of modernizing of this ancient point of view, of course, especially since the middle of the twentieth century. For example, the German school of thought called "philosophical anthropology," which studies the human-technology relationship, argues, broadly, that far from being distinct from technology, we humans *are* embodiments of technology. We could not have survived as a species without evolving with and through technological invention. It's an idea developed more recently by a growing range of thinkers in disciplines ranging from computer science to sociology, and which argues that we are in fact "cyborgs"—part technology and part organism. The cyborg is the evolving human integration, taking place over many hundreds of years, with gradually more complex forms of technology. From the use of false teeth, devised by the Romans, to the invention of spectacles, which dates from the 1300s, and from the pacemaker that regulates the heart, first implanted in the 1950s, to the Wi-Fi signal that enables access to networked digital technologies, the body and technology have been assimilating for a very long time. Such evolution was a process of *interaction* as well as integration in the view of someone like Marshall McLuhan. He wrote influentially that the relationship is one of mutual constitution. That is to say, the technologies we create also shape the ways that we think and act in the world, and this, in turn, influences the kinds of technology we further invent and use. And on the process goes, in a historical-evolutionary cycle.

The technologies we create also shape the ways that we think and act in the world, and this, in turn, influences the kinds of technology we further invent and use.

Then there's *loss*. What we lose because of participation in digital life is the supposed "realness" or "authenticity" that comes with analog technology. On that point McKibben and Sax coincide. And there are others, too, who say that with the rise of digital, something has gone from our lives, something that the phoniness of digital virtuality readily expresses—"There is no there, there,"[2] as cyberpunk novelist William Gibson put it. Analog has become, without our asking for it, a thing outmoded, superseded, retired, or just *lost* to our daily experience, something replaced by a much less fulfilling technology. We're told that collectively and individually digital has given us so much that enhances our lives, but there's a voice in our heads, or just a feeling in our soul, that tells us that we've suffered a psychic forfeiture in the rapid transition to days spent looking at a screen.

All this talk of the "loss of the real," however, is rather vague. And that's our first problem. Analog presents itself to us today as something—an experience or process—that's hard to put a finger on. And because we can't properly identify it, we call it "nostalgia" or "retro." Which doesn't tell us very much. Nevertheless, we blithely create such consumer experiences for ourselves, of the kind that we see in the fads for new (or old) analog merchandise that seem to appear in the stores or somewhere online every week. Whatever marketers and savvy businesses can dream up can become the next nostalgic trip down memory lane. This could be *anything* that can plausibly be labeled "retro," such as flimsy cassette tapes, or a kitsch 1950s wall clock, or "quality" Moleskine notebooks that evoke (according to the manufacturer's website) the analog technologies of choice for Picasso, or van Gogh, or Hemingway. You don't even have to have been born in a predigital time to crave the retro experience. Anyone can be gripped by the urge to purchase those reissued 1985 Air Jordan 1 sneakers, or the Fujifilm Instax "Classic" analog camera, just to get the analog fix to fill that analog gap that exists somewhere inside us.

The title of McKibben's piece in the *New York Review of Books* is "Pause! We Can Go Back!" This is perhaps the mordant humor of the editor. But it is a story of our digital age that there is no pausing, just as there can be no going back. And neither can we go forward, positively, without an appreciation of *why* we would wish to pause or go back anyway.

So we need to know analog better. We need to be able to identify the analog gap in our lives that comes from an inability to clearly understand the human-technology relationship, especially now, in the context of domination by digital. We need to know, for example, what's going on when we read about the audiophile Pete Hutchison, from London, who spent thousands of pounds scouring the world for the last remaining original parts of analog recording equipment, mostly 1950s and 1960s vintage, so that he could produce totally original-sounding and original-looking re-pressings of vinyl records from the "golden era of classical recording." This is a wonderfully romantic idea that seeks to be true to an analog ethos in a way that most vinyl records today (pressed from digital master copies) aren't. Nonetheless, market economics will dictate that a box set of Mozart's complete "Parisian" works on seven discs, conducted by Fernand Oubradous in 1956, would cost you around US$3,500 from Hutchison's Electric Recording Company. Hutchison says, "It's not about vinyl . . . it's about a whole philosophy."[3] Indeed. At the time of writing, every single one of Hutchison's fifty or so lovingly recreated LPs, with exactingly replicated sleeves and disc labels, from Beethoven to Bruckner, and from Sonny Rollins to Thelonious Monk, and in strictly limited editions of three-hundred copies for each, are listed on the company's website as "sold out." Analog philosophy is also very good business.

Why people like Hutchison do this and why people respond to such sentiment is important to grasp, because the relationship to technology and technological things, analog or digital, is at the very center of our being. I wrote previously about analog "fads." Well, if we understand fads to be temporary phenomena, then I suppose this fits in respect of the specific commodity, which will come and go. But whether it's Coltrane from 1957 on the Prestige label, or a 1960s Astra lava lamp, is not the point. The point is that the analog urge is deep-seated in individuals and societies. It's about something lost from our evolutionary past with technology, something not there in our digital present. And it's about us, always trying to find that "something" in products that, sensing our need and our urge, capitalism readily fulfills.

In other words, we need a better awareness of where analog sits within individual and collective consciousness—and today that's mostly *obscurely*.

***

Did you know that notwithstanding our undiminished taste for analog technology and retro experiences, the word "analog" itself has been steadily disappearing from our language? I know this because Google has a program called Ngram that can search the millions of books it has scanned since 2002. It can look for the frequency of words in its vast database of texts that date from the 1500s until 2019 and convert this data into a graph to show the word frequency over a very long timescale. "Analog" began to appear in English language books around 1800. The Ngram graph line for this word runs steadily across the bottom of the chart and remains there until 1947. Then the graph line shoots up at a steep angle until it reaches its all-time peak around the mid-1980s, whereupon it drops as sharply as it rose. Interestingly, the trajectory of "analog" mirrors that of "digital" almost exactly. Appearing in print around the same time the words travel in parallel until the late 1940s when "analog" begins its decline—but digital carries on up in its steady ascent until 2019—when Google abandoned the project.

The path of "analog" in the Ngram plot tells us something. It tells us that as a word declines in frequency in our print culture, it declines also as a part of our language, written *and* spoken. It follows that its decline in language means the *decline as a concept*, as an idea, as a recognized component in the meaning-sharing that written and spoken communication sustains. The result is that as analog begins to disappear as a mode of knowledge, it therefore is even more difficult to comprehend.

The human relationship with technology is our very earliest connection. It is what made us *Homo sapiens*, the self-proclaimed "wise" creatures who evolved with tools to dominate all other species. And until the emergence of digital these were always analog tools. And so, the hankering that might suddenly oblige someone to go online and purchase a "746 Rotary 1970s-style Retro Landline Phone—Curly Cord, Authentic Bell Ring—Mint Green," for $86.99 at the BigaMart website, needs to be understood as more than an impulse buy of a piece of overpriced retro. It might be better seen as a desire to plug that gap in our psyche through the gesture of consumption. But as with the consumption of most commodities, it never

The human relationship with technology is our very earliest connection.

really satisfies. So we keep coming back for something else that we see: a pair of 1940s 501 Levi's from the Beyond Retro website that will again strike that mysterious chord of emptiness within us. We may buy it, or we may not. But what we don't do enough is to properly interrogate the feeling within.

Investigate and reveal is what the following chapters will do. And then the faint wish that may cross your mind one day to have a beautiful and dependable brass key to open the door of your home or workplace instead of an unconvincing plastic swipe card may well go unfulfilled in these days of digital, but at least the desire and its source will be more understandable.

The happy *frisson* we may feel when we see, touch, or use an analog technology is not just nostalgia, but *recognition* too.

2

BEING ANALOG

> The distinction between digital and analogue representation is philosophical before it is technical
>
> —Chris Cheshire, "The Ontology of Digital Domains"[1]

This is a book about *analog*. Primarily, it's a book about analog technologies and our historical relationship to them and how we have used them to shape our world and our cultures and societies in certain—extremely important—ways. But it's about more than that. It's analog without the definite article, which suggests, or is intended to suggest, a wider-angled focus and a longer depth of field that will reveal unexpected things. And these are things that come into sharp focus when set against digital technologies—and are unexpected in ways that go beyond the clichés of analog as simply signifying the past and the outmoded, and digital as epitomizing a shiny present and glorious future.

Analog is something that I've wanted to write for some years now. What the following pages contain are the combed-out tangles of several years of notes, of published writings, of recollected conversations, of shelves full of books on analog and digital technology and the cultures they produce. I draw also from a large personal database of photographs, Internet bookmarks, video files, interviews, and films. The book is a distillation of all these into what I see as the essential knowledge of a (still) essential subject. And so, part of what I want to do here is to show that essentiality in that wider-angled context. This means not only a mini history of some of the major technologies and epochal phases of analog technological development and their immense importance to human survival and expansion, but also how this history and this association has tremendous relevance for us today as we try to understand and navigate in a world dominated by *another category* of technology, the digital.

Our age is indeed digital. Part of what this means is that the look, texture, and our tacit knowledge of many analog technologies, such as the hook and dialer of a big rotary telephone, are receding from working memory as digitality encloses ever more spheres of human activity. This also means that for generations born after the turn of the twentieth century, memories of when the home-based fixed-line telephone was all there was for this type of communication will be nonexistent. Nonetheless, it's interesting to note that such memories, real or imagined, of the telephone on the kitchen wall, or the rack of vinyl albums next to the stereo, or the cumbersome television set occupying the most dominating place in the living room, are usually fond ones either based on personal experience or generated as nostalgia-by-proxy from watching perennial reruns of pre-1990s TV shows or films that depict a comparatively recent predigital life. Selective they may be, but analog memories contain an undoubted *lack of antipathy* toward the technological form and function they represented—and this tells us something important.

It is often assumed that analog technologies disappear from our lives because they are not quite up to the job anymore when tested against its digital challenger, although this perceived deficiency always seems to be measured in terms of nebulous concepts such as efficiency and progress. But it would be a mistake to accept such change merely in its own terms and for its own sake, even when the change is seemingly trivial. Think of the once-ubiquitous analog needle meter. This rather unexciting contraption has nonetheless a venerable lineage, deriving from ancient sundials and later numbered clocks where the creeping shadow and the hour/minute hands indicated or "read" the passage of time. In its own time, the twitchy and humble, usually spring-driven needle became a way to understand elements of lived social experience by reading or measuring all kinds of things in everyday life, such as temperature and barometric pressure on a household thermometer, or voltage meters, or gas flow readings, or the sound volume on the home stereo system. Indeed, the stereo meter with its eardrum-splitting red zone was iconified in pop culture as the volume unit (VU) display on the cover of the Velvet Underground's *VU* compilation album from 1985.

Like many analog technologies, the needle reader is becoming a curio for the collector of vintage equipment or, in the case of the VU display, a superfluous decoration on expensive stereo systems for audiophiles who still like that kind of thing. And of course, for an example of far more durable analog technology there is the wristwatch, with its automatic (*never* quartz-digital) movement, which is still the horological marker of quality, distinction, and discernment over the cheaper and ironically more accurate rival.

Mostly, however, analog forms and functions dwindle from use without our noticing, although sometimes we do notice and are glad about a certain technology's passing. For instance, analog television was universal not so long ago. But its sinewy and continuous analog signal meant that viewers were at the mercy of any obstacles in the way of its transmission, such as unsettled atmospherics, the changing built environment, rooftops, trees, hills, and other obstacles. Such were the variables that could play havoc with the wavelength signal and were a constant threat to interference-free viewing. By contrast, digital signals are short and discontinuous and therefore supposedly more robust. But these too can be blocked by mountains or strong winds that can cause the screen image to pixelate or the signal to drop out completely. The distinction is interesting: the digital picture is either "on" or "off" due to the obstruction; you either can watch or you can't. Digital does not come to you in some half-watchable "in-between" state as it does with analog. Analog's wavy and fuzzy indeterminate state offered scope for human physical intervention in a way that pixilation doesn't. Armed with a movable antenna the putative viewer could try to catch the unruly signal and hope to find a sweet spot that would steady the image and soothe the angry white noise. With the digital signal, there is only a helplessness, a disconnect not only with the signal but with your relationship with the technology. Nevertheless, not many who underwent the sometimes-daily analog duel with the TV set will miss it. And with analog TV all but phased out in many countries, it is a memory that dwindled, unlamented. As we will see, however, the physical interaction with technology runs very deeply in our evolutionary history, and this tells us something elemental about analog and about us.

Notwithstanding the disappearance of many analog machines and processes, deeper cultural and psychological memory traces still flicker in the mind's eye like a grainy picture on an old TV. These can reveal themselves in countless ways in our psychological lives, such as realizing one day that the box of analog VHS tapes in the basement, the ones containing washed-out scenes of family birthdays and holidays and of Grandad when he was still alive, are going to be difficult to retrieve now. The expensively purchased Akai videotape player was thrown out years ago. Nostalgia and sadness mixed with a cultural passivity regarding the supposed inevitability of technological change in the service of progress can provoke an emotional pang of loss, the loss of something beyond images on a fragile and degradable oxide tape. It's a loss of a memory represented as an image, but it's also the loss of cultural memory embedded in the technologies that created them, images locked into obsoleted media, images there to decay and eventually to die.

We can see a more general example of this felt cultural loss of analog forms in Michel Hazanavicius's film *The Artist*, from 2011. It is set in the transition period in moviemaking from silent to talkies between 1927 and 1932. It's filmed in sumptuous black and white using Kodak Vision3 film stock. The cinematography is lovingly observant, depicting the existential crisis experienced by the protagonist, silent movie star George Valentin, when one technological form, the one Valentin thrived on, gives way inexorably to another. Hazanavicius had difficulties in securing funding for the film. Hollywood executives could see no commercial interest in a silent film shot in black and white. However, the film was a triumph, achieving much critical acclaim and well as box office success. The film is beautifully shot, scored, and acted, but more than anything, what drew the crowds and the critics to *The Artist* was the tapping of a deep-lying well of analog nostalgia that exists in us, something that can be released by the reenactment of analog forms and worlds in celluloid—even though these were outmoded long before most of its viewers were born.

The Artist makes much out of the power of nostalgia and draws liberally from cultural and technological history to represent the sometimes-inexpressible categories of loss in the face of constant change. However, *Analog* is not a lamentation along the lines of "old is good" and "new is bad." Nor does it lean wholly on the power of nostalgia to make its point. That would be uninteresting and uninformative. Rather, the book reveals the human bond with a particular category of technology, analog, and makes its importance apparent when set against a different category that has eclipsed it, digital. In other words, until quite recently it made no sense to think much about our connection with the analog beyond a vague feeling of nostalgia, because prior to the rise of digital, analog was on its own, with no challenge to its domination. Now there is. And with that stark contrast we can begin to appreciate the principal features of the long eons of predigital life more fully.

Figure 1 Unashamed nostalgia: poster for *The Artist*, 2011

***

Words matter. This is a cliché. But in the case of analog, it is one that is too important to ignore in respect of what the words' etymology can tell about meaning and how it changes through the ages in a process of cultural evolution. In terms of meaning, *digit* or *digital* is straightforward, with ancient roots in the measure of the human hand to represent a unit of length, a definition that was mathematical and rational from the beginning. Analog and its meaning(s), by contrast, are rather more *baroque*: there is no simple and suitable and stable definition that may be universally applied. What we find, rather, is the irrational, the imprecise, the ineffable, what Kant called "the crooked timber of humanity"—everything, indeed, that corresponds, in an unpredictable and irregular way, to what we are. The allusion to humanity is a clue to understanding a key aspect of analogicity.

In 2016 Jonathan Sterne wrote an essay, "Analog," for a collection called *Digital Keywords*.[2] It is one of the few recent works that seeks to explain and contextualize the term in our digital age. However, it tends to reinforce a contemporary understanding of what analog means—which is to equate solely it with a technology. But if we take a more historical-etymological approach to the word, then we find a more inclusive and expandable definition of "analog" that goes beyond the current meaning to reveal an idea and a process that is social, cultural, historical, *as well as* technological.

Before getting to that, however, it's instructive to relate how analog is defined in a simple Google definition search, something many would resort to today to get a handle on the word. Google gives an adjectival definition:

> adjective: **analog**
>
> —relating to or using signals or information represented by a continuously variable physical quantity such as spatial position, voltage, etc.
>
> "analogue signals"

Such is the humdrum definition you might expect if wanting to check quickly online or in the concise commercial dictionaries; it echoes much of the popular cultural discourses about analog. It refers to *technology* and could easily map on to the discussion above on needle meters and TV. The etymology is more wide-ranging than the electrical signal or the vinyl record. The *Oxford English Dictionary* (*OED*), for instance, tells us that "analog" has "multiple origins" and these have generated what it calls the historical "general senses" of the word, which strongly implicate a human factor.[3] In other words, if we look just a little more closely at the "multiple origins" and "general senses" of the term, we find meanings and characterizations of analog that take us to a whole new world.

Digging into the etymological record it is remarkable to see that the more ancient meanings of analog immediately give a human dimension, an embodiment, to the word—something absent in the popular retro meanings today. "Analog" has its antecedents in a group of related Greek and Latin words, *analogon*, *analogate*, *analogous*. These are part of the "multiple origins" that feed into the French *analogue*, which means, according to the *OED*: "A thing which (or occasionally person, who) is analogous to another; a parallel, an equivalent."[4] Parallel. Proportionate. A person or thing. The human and the tool, the human and the world. Straightaway, the contemporary definition becomes more expansive. The exciting discovery from a little research is that a person is (or can be) an analog.

We can delve a little deeper. These meanings are the *OED*'s "general senses" of the word that denote a connection in which—although not necessarily so—the human is implicated in the proportionality, the equivalency, the paralleling and the general relationship or relatedness that these ancient root-words convey. Entries in Samuel Johnson's *Dictionary* of 1785 are "analogal," which he defines as meaning "having relation," and "analogous," which is specified as "bearing some resemblance and proportion; having something parallel."[5] Nothing remotely technological here. Analog is an interaction between things, humans included, integrated and evolving in all of Johnson's several related entries. Interestingly, Johnson's entry "workmanship" in the same dictionary gives another fascinating insight into the understood general senses of analog as it relates to labor, the human body, and, by implication, technology. He notes:

> Workmanship; power or act of manufacturing or making.
>
> An intelligent being, coming out of the *hands* of infinite perfection, with an aversion or even indifferency to be reunited with its Author, the source of its utmost felicity, is such a shock and deformity in the beautiful *analogy* of things, as is not consistent with finite wisdom and perfection.[6]

In early modern times, such general senses and multiple origins constituted a more holistic attribution in that they attempt to capture the widest possible scope of human interaction with technology and the transformation of their environments with it. However, at least since the emergence of digital technologies in the 1940s, we stopped using these wider definitions and became acquainted more with the idea of analog as somehow residing in or defining of the tool or machine or apparatus itself, and only then as the flip side of that which rose to replace it—digital. The general sense became a particular sense: a technological sense. Earlier meanings and definitions began to disappear—if not from the older and more comprehensive dictionaries, where we can still find them, but certainly from colloquial speech and contemporary writing.

There is a need to rediscover the interaction between the anthropological, natural, and technological elements of analog as human activity in the world so to give a better understanding of analog's place in the context of our digital-dominant reality today. The Greek and Latin roots of "analog" are a good place to begin. They reflect an ancient, but still limited perception of the human link with technology. Indeed, philosophers such as Democritus and Aristotle laid down an important marker of understanding that in retrospect gives insight into why the human component of analog becomes diminished in more recent times. It was the pre-Socratic Democritus (460–370 BCE) who first cogitated the qualities of technology and the position of the human in relation to it. He gave us the idea that "technology imitates nature." This is the notion that the very earliest humans looked to nature for ways to survive and thrive through technology. For Democritus, the skill of the spider in the action of "weaving and mending, or the swallow in house-building," are examples of the cues in nature that instructed humans to adapt and learn to control their environments. Significantly, Democritus implies an already-developed human intellect that could stand back and look at nature and reflect on its usefulness for an already-underway human project of development.

It is an argument that Aristotle (384–322 BCE) agreed with. Moreover, he explicitly suggested that "imitation" implied a separation—human from technology—that posited an ontological division between natural things (humans, nature) and artifacts. Democritus and Aristotle were thus responsible for the very big claim, one that hung on through the centuries, at least at the commonsense level, that humans exist *apart* from nature and that their technologies are simply imitations of it. And so analog, in this foundational view, applied more to the general sense of equivalency instead of something more intimately human.

We get a much more human sensibility from Arnold Gehlen, a contemporary of Martin Heidegger of the German philosophy of technology school. Gehlen wanted to take the early philosophical debates into another arena and up to another level of inquiry. He wanted to think about the separation question not only from the perspective of philosophy but also from that of anthropology, or the study of man in the past and present. This more inclusive and holistic approach came to be known as "philosophical anthropology." In his *Man in the Age of Technology*, Gehlen argues that in our present evolutionary state, and in our present physical and cognitive state that stretches back 200,000 years, we are born "unfinished"—underprovided-for beings who are "poorly equipped . . . with sensory apparatus, naturally defenseless, naked, constitutionally embryonic through and through, possessing only inadequate instincts."[7] For Gehlen the human drift toward technology was vital for our survival and, once established, acted as the link between "man and his organic and instinctual deficiencies and a hostile natural environment." We *evolved with* technology to become *Homo sapiens*. We did not "discover" it. We are still born "unfinished," and the newborn will die without nourishment and care. But we have developed as creatures who, unlike other species, can survive through technology—or *as* technology in continuous interaction with our environment and the material world.

This oneness with technology and nature occurred hundreds of thousands of years ago. And it was during that unknowably harrowing evolutionary struggle for life, lost mostly to the fossil record, where countless mutations and adaptations in our biology over numberless generations led to a point when our earliest species' capacity as tool makers and users reached a level where we could begin to survive. At its most elemental level, this connection finds expression in what Gehlen terms the "circle of action," an ancient—perhaps *the* most ancient—interactive process where, as he put it, the relationship with technology "goes through object, eye, and hand and which in returning to the object concluded itself and begins anew."[8] He continues: "The analogous process of the external world

We *evolved with* technology to become *Homo sapiens*. We did not “discover” it.

bespeaks a 'resonance' which conveys to man an intimate feeling for his very nature, by focusing on what echoes his nature in the external world. And if we today still speak of the 'course' of the stars and of the 'running of machines,' the similarities thus evoked are not in the least superficial; they convey to men certain distinctive conceptions of their own essential traits based on 'resonance.' Through these similarities man interprets the world after his own image, and vice-versa, himself after his image of the world."[9]

In these sentences the term "resonance" plainly resonates with Gehlen. He chose the term carefully. Resonance is from the Latin *resonantia*, meaning "echo" or "vibration." This relates to the general senses of analog. It suggests a sensual and physical connection, as in when something or someone "resonates" with you—for example, the "vibe" that someone may emit, which can be positive or negative, but there is an unmistakable rapport, and a potential union if the resonance is shared and felt at some level. This is what Gehlen was getting at: a oneness, or an "intimate feeling," or understanding, that places humans at the very center of technologies composed by themselves of natural elements drawn from their immediate environment, such as stone, wood, and, later, durable metal.

Gehlen's book was published originally in the German in 1950, before the full-fledged emergence of digital, and a time when analog, and our relationship to it, was a question not readily presenting itself as one to ask. Moreover,

Gehlen was concerned not so much with the question or problem of whether humans themselves are analog, but with establishing our unity with technology and the "external world" of nature.

Today, more thinkers are giving sustained attention to the general senses of analog and what it suggests for life in our expanding digital context. Silvia Estévez, for example, is an anthropologist who studies human migration and technology use and has opened up valuable insights into how the general senses of analog can allow us not only to see the world differently in relation to digital but to see ourselves differently too. Estévez tells us that humans are analog—"analog human beings"—and that we don't mix well with computers. Quoting the computer scientist Charles Petzold, she writes: "people and computers are very different animals."[10] To illustrate the distinction, Estévez builds on the general senses to make some fundamental points about what "equivalence," "proportionality," "parallel," and other analog definitions say about us and our relationship with technology.

"Analog machines," writes Estévez, even complex ones such as trains or ships, have operations that "simulate processes that people had seen before in nature and in their own bodies."[11] This situates people directly into the technology process through our correspondence and resonance with them. For Estévez a key characteristic of analog technology is one whose "activity crosses time and space in a visible way that allows us to grasp the link between a movement and its effect, the process, the continuity."[12] In other words, there is a recognition of the action of an analog technology, wherein we can see, and to some degree understand, what it does and how it does what it does. We recognize what a technology does because "the process, the continuity" is iterative. The automobile, for example, recognizably builds upon the buggy and horse, and tracing further back, the horse itself (as a technology of mobility) we recognize as an improvement to our own fundamental capacities for mobility through walking or running. Compare this with digital processes generally: most of us are unable to adequately recognize how time and space shrink through digital networks, and indeed how physical space becomes virtual. This is because, being a different category of technology, "the process" is different in respect of the techno-logic. As Gregory Bateson, an early theorist of digitality, argued, an analogical system is continuous, while a digital system is discontinuous.[13] There is no "continuity" in respect of digital's logical activity and nothing in respect of a historical process of iteration that we can trace back to more ancient human technological and anthropological roots.

The ideas of Marshal McLuhan give this holistic analog story some more heft. McLuhan was that unusual thing: a philosopher of technology whose ideas captured the *zeitgeist*. His *Understanding Media* was a popular sensation. It was lapped up by millions of readers who rarely had been given such learned and powerful perspective into the everyday association with technology in a mass-mediated society. McLuhan's ideas are a staple today in high school and university media and communications subjects, and his celebrated aphorisms provide a good shortcut to his deeper ideas for a wider readership.

Two "McLuhanisms" stand out in relation to analog. One was his concept (mooted over a decade earlier by Arnold Gehlen, incidentally) that "Any technology is an extension of our physical bodies." Technology becomes an extension or, as he termed it, a "servomechanism," whereby the "[Native American] is the servomechanism of his canoe, as the cowboy of his horse or the executive of his clock."[14] Technologies, in other words, are the "amplifications of our own beings" with, for example, the ancient flint knife acting as the extension and intensification of the actions of human teeth or nails, or the book as an amplification of cognition, or spectacles as an amplification or extension of our visual capacity. We see here, too, Estévez's criteria of recognition, where the process and the continuity of such amplification or extension is iterative and clear. This process can scale up to incorporate a great deal of analog complexity. And so, as technology develops, the modern airplane, for example, not only is the amplification of the actions of our bodies in space through mechanical flight but also is an "imitation" of flight in nature.

Human mastery over technology in this respect achieves what Gehlen called *replacement technique*—technologies such as airplanes or submarines—that act in place of organs or capacities not naturally possessed by humans.

McLuhan's other dictum, "the medium is the message," likewise encapsulates a profound insight into the human ties with technology. In *Understanding Media*, he defines it unambiguously: "The personal and social consequences of any medium—that is, of any extension of ourselves—result from the new scale that is introduced into our affairs by each extension of ourselves, or by any new technology."[15]

It is the *medium* that is most important. What the technology, always as an extension of ourselves, enables the individual or society to do is the key factor. This is where *potential* and therefore technological *development* lies in our species' interaction with analog technology. This insight, which will be an underlying feature of the rest of the book, is that the real message is the potential inherent in the interaction with technology that places the human dimension at the heart of the process. It is a dynamic or motivating force that for 50,000 years has acted against inertia, keeping human-technological interaction moving to solve human problems, satisfy human curiosity, and almost always to provide the "new scale" potential that meant increasing human capacity and increasing technological innovation.

***

It's worthwhile recapitulating where we are before moving on. To know what we are thinking *about* when thinking about analog is important if we want to get the most from an essential knowledge perspective. And what constitutes "analog" in its wider historical significance is something we all should know a little more about. However, that the status of analog has gotten so little sustained attention in recent years is in one sense testament to the rapid retreat of the analog way of life by computerization. A reason for this might be that we have never really had to examine too closely our technological relationship with analog because our predigital technological relationship was monogamous: analog was our only partner until very recently. There was nothing with which to question its relationship to us, and us to it. Analog as a process, a relationship, and a technology has not gone, of course. It's just that digital technology has abruptly colonized modern life in so many spheres that we have never really asked ourselves properly what is the obvious question: What is it that has just been overtaken as the primary techno-logic that governs much of our collective and individual lives?

Easily accessible *OED* sources clearly implicate human involvement in the understanding of the meaning of analog. And it is in this incorporative sense, drawing from a wide range of knowledge disciplines, that I define analog; from the Greek and Latin, from the lexicographer Samuel Johnson's taxonomies of knowledge, from philosophical anthropology and social anthropology, and from environmental philosophy and cognitive science, I explain the ancient analogical interaction between human, technology, and nature. It was an interaction that McLuhan understood in *Understanding Media*. And it was one defined beautifully by John Culkin in a 1967 review of McLuhan's book, expressing it thus: "We become what we behold . . . we shape our tools and afterwards our tools shape us."[16] And, he may have added, this analog connection draws from nature for its material and reconstitutes the natural environment through constant innovation, adaptation, and the technological expression (as an aspect of the relationship) that has taken infinite forms throughout human history.

The contemporaneous eliding of the aspects of humans and nature from the analog definition blinds us to the variety and diversity of meaning that it has contained historically in Western culture. By construing analog only in its technological sense, we reinforce what Sterne called the "common sense" definition that inevitably misses the depth of meaning and history of human experience with technology and through the natural world. It obscures what has in fact been a symbiotic and adaptive relationship that constitutes more than we currently see. But as we will see, to understand ourselves as being a part of the analog process, indeed, to *be analog ourselves*, not only places us properly at the center of our technological history but shows that history to be not just about the evolution of technology—but also about the evolution of human beings who are both creatures of technology and technological creatures.

3

RETRO ANALOG (THE ZOMBIE IN THE DIGITAL MACHINE)

We inhabit a world of digital communications where most of us could live pretty much all of our lives online. The global pandemic of the early 2020s tested this assumption, and it proved to be true. The health emergency forced many millions of people into lockdown and physical distancing measures that moved us away from each other and closer to our computers and mobile phones for virtual friendship, solace, entertainment—and work. It was a proof of concept that Nicholas Negroponte, founder of MIT Media Lab, prophesied in long-ago 1995. He wrote that we were almost literally *becoming* digital and that with new computer technologies becoming omnipresent, bytes would merge with atoms in the construction of a new kind of human. A by-product of this technological transformation, he noted, was that analog technology such as television and the vast ecosystem it had built up over the second half of the twentieth century would be eclipsed as obsolete, and even the ways that we thought about technology would be consigned to the "old-age home for analog thought."[1] We were compelled to adapt, and the pandemic activated an explosion of Zoom-led virtuality that showed that a great deal of life online was actually possible, if not necessarily preferable.

In truth, we had been heading this way for a generation or more. The "information society" was a recognizable "thing" by the early 1970s, with Daniel Bell telling us, correctly as it turns out, that we were becoming "postindustrial," meaning we were more "knowledge workers" as opposed to machine operators.[2] Western societies were becoming centered more around the production of services and computers provided the (literal) link between knowledge, services, and a changing economy and society. The sociology and psychology of the change from working with material things to working with information of some sort was noticed by more than just a narrow band of specialist academics and economists, however. For example, Robert Pirsig seemed to touch a deep chord with his 1974 book *Zen and the Art of Motorcycle Maintenance*—an instant hit, never out of print to this day, and apparently the most widely read philosophy book ever. It's a road trip book framed in a personal philosophy of "quality." Pirsig rhapsodizes about life outdoors and on the road: about wilderness, mountains, rivers and sky, and such things as stopping his motorbike under a shady tree somewhere in America's Northwest, to tenderly tune its engine, tighten the tappets, and caress the porcelain parts of the spark plugs as he cleans them, for over four hundred pages. Many who had never ridden a motorbike like Pirsig's 1966 Honda Super Hawk nonetheless saw in the writing an elegy for a seemingly more authentic and tactile age. And it was sign of the appreciation of the importance of our connection with such machines that the Smithsonian Museum purchased Pirsig's Honda in 2019.

Of course, analog machines, including freedom-giving motorcycles and automobiles, did not disappear overnight. In a process of "remediation," many just became digital to a greater or lesser extent. The typewriter, for example, morphed wholly into a *word processor*. The vinyl disc was usurped by the compact disc (CD) to become a different thing doing the same job but with different effect. Superseded telephones, televisions, newspapers, cameras, film stock, and indeed the analog computer itself either adapted to the new logic, checked in to their old-age homes as semi-obsolete (such as the analog clock), or faded away due to their lack of capacity to compete with the digital rival (video and audio tapes, AM-FM radio, vacuum tubes, letterpress printers, etc.).

Analog machines and process remain with us in many spheres of life, and still perform many important functions. However, the problem for cultural memory is that they are overshadowed by a digital logic so all-pervasive that they are consigned to the subconscious, or they disappear from use and view altogether. It is this that feeds the feelings of nostalgia and lack that we saw in the previous chapter. Expressions of this are seen welling up from what has become an analog subculture. Audiophiles speak of the "warmth" in the sound of a tape machine, or vinyl record, something stripped out from the "cold" tones produced by the industry-standard MIDI computer file that is burned into a CD. Similarly, the guitar and vocals of, say, Leadbelly, or the middle-register soprano of Maria Callas, the argument goes, is better on vinyl than on CD. A vinyl recording, the smell of a new printed book, or even the crooked unpredictability of the sinewy long-wave radio signal finding its way to the analog receiver are said by many—and perhaps unconsciously thought to be so by more—to be something *authentic*. And authenticity is a powerful driver of the cultural phenomenon of *retro*, a term of 1970s coinage that has a life and place of its own in the essential knowledge of analog.

***

The somewhat brief life of the Crosley Corporation of America stretched from 1921 until 1956. This was a period that roughly corresponded with that of a kind of high

Authenticity is a
powerful driver of the
cultural phenomenon
of *retro*.

modernity in the United States, a golden age of Fordist mass production that had propelled America to the very top of global superpowerdom. Crosley Corporation was founded by Powel Crosley, "the Henry Ford of radios," and was headquartered in Louisville, Kentucky. Alongside radios, it specialized in the manufacture of automobiles. In the 1950s, with television just hitting its straps as another big consumer technology, cars and radios were still the iconic staples of postwar Americana. And Crosley itself was a mini-icon of its time. The company name, the period, and the products conjure in the mind's eye a Norman Rockwell world of material plenty and Anglo-Saxon confidence and contentment.

Strange, then, that Crosley Corporation went bust when it did. In the year 1956, the combined might of American auto manufacture produced Fords, Chevrolets, Plymouths, and other cars to a total of 6,203,027 vehicles; it was also the year that Elvis Presley's eponymous debut album and singles "Don't Be Cruel" and "Hound Dog" sold millions of vinyl copies. Being a pioneer in auto disc brakes wasn't enough, and being the first to introduce push-button radios in their cars didn't save Crosley. That their "Hotshot" sportscar and "Pup" radio would one day become valued collectors' items likewise failed to shield this quintessentially small-time rural company from the perennial hazards of the big city corporations' market power, even in boom times.

Odder still is that the brand name Crosley was disinterred by a Louisville marketing firm, and in 1992 the new company began making vinyl record players in vintage styles from the 1950s, 1960s, and 1970s. Odd because this may have been something of an inauspicious period for such a venture in music industry. Vinyl record sales were already plunging as an effect of the introduction of the digital compact disc a decade previously. And the allegedly unscratchable and unbreakable CD would dominate music buying until around 2000, when it began to give way to Napster-type online pirating and iTunes buying and, later, streaming services like Spotify. Reviewing its bold move, the maker's website, CrosleyRadio.com, recently announced: "Our first turntable was released in 1992, when CDs were still king. Naturally, everyone thought we were nuts. But now, after over 30 years of bringing stylish music to the people, we're one of the biggest manufacturers and trendsetters of the new-millennium Vinyl Resurgence."[3]

Reviews of its products were mixed, notwithstanding the website promotion. Reddit thread observations on its turntables, for example, range from "pretty bad" to suffering "random skipping issues." And commenting on the Crosley contribution to the Vinyl Resurgence, the *Guardian* music critic Alex Petridis noted that the turntable he purchased for his daughter was of "shoddy build" and that the tone arm had to be weighed down with a coin to stop it skipping. Petridis softened the criticism by saying that

Figure 2 Screenshot from crosleyradio.com

they did "look like fun, in a way that hi-fi equipment seldom does. They don't need setting up: you just plug it in and it plays."[4]

And it must be said that the magnetism of such technology can't be denied. But for many an adult old enough to remember, fun combines with a nostalgia that is leavened by the recollection that except for high end hi-fi systems, many if not most predigital mass market record players were cheap and of poor quality, and the Sellotaped coin on the tone arm was a "fix" that many had recourse to. The analog resurgence that the Crosley website speaks of is not confined to turntables. Its retro products sell into evidently healthy consumer markets for jukeboxes, wall phones, cassette players, and hi-fi shelf systems. And Crosley is not alone in a music-based analog market that has seen the reappearance, to take one instance, of reel-to-reel technology for tape music at the connoisseur end, with very expensive machines made by the Über-cool German manufacturer Ballfinger. And at the other end of the tape market, there has been a mass-consumer revival of 1979 Sony Walkman–type cassette players, such that you can buy them in supermarkets, with a crackly AM-FM radio included.

All this is of a piece with a broader renewal under the retro rubric. David Sax's book *Revenge of the Analog: Real Things and Why They Matter* tries to capture, and understand, retro in a readable and enjoyable journalistic panegyric.[5] The main title sounds like a publisher's suggestion, a catchy line that echoes a 1950s sci-fi B-movie. The subtitle is more interesting and is likely what motivated Sax to write it. The "Real Things" he refers to harkens back to a mythic past in our collective Western cultures. Again, this conjures a Norman Rockwell time of Barbie dolls, of model train sets, and the nuclear family, a time when food tasted better and was more healthful. It was an imagined time when machines were useful and met a need. A time when machines and products were built locally, built properly, and built to last.

Sax's book covers some of what you might expect, such as the surge in vinyl popularity from which the Crosley company presumably benefits. The US vinyl industry, he tells us, was almost extinct by 2007 when shrinking demand meant that its few working vintage presses stamped out only 900,000 albums that year, way down from an early 1980s zenith of over 300 million. However, old-school record buying bounced back from this historical nadir to sell over 12 million copies in 2015. More recent analysis confirms that vinyl sales have overtaken CD sales for the first time since 1986. But Sax speaks to something more than the (usually) black discs made from an oil-based compound of polyvinyl chloride. It is a mysterious thing never fully explained but is expressed as an irrepressible hunger, a nostalgia, a curiosity, or a human need for the "real things" of his subtitle. In parts Sax gets almost metaphysical, writing, "Analog experiences can provide us with the kind of real-world pleasures and rewards digital ones cannot."[6] The more that digital permeates life, the more, he argues, that people make the effort, in both time and money, to seek some sort of consolation from something that triggers a resonance somewhere inside us. Nowadays, even the post-boomer generations are seeking out analog encounters for themselves. The younger culture vultures are driven not by a quixotic attempt to recapture some fading memory trace but by an innate curiosity about an alternative category of technology with a different look and feel from the digital ones they had been weaned on.

Business journals and financial supplements lead the journalism on the analog renaissance, for the obvious reason that it constitutes new businesses and new industries. Sax devotes an entire chapter to the revitalization of board games. What he terms the "revenge of the board games" is a cultural craving for the "real" that generates a $10 billion industry. It's a business built on traditional games such as Monopoly and Connect-4, games that had never really gone away. However, it also includes games that were born digital and virtual but have been reincarnated as cardboard and plastic "real things," games such as World of Warcraft and StarCraft that sell in huge volume across a wide demographic spectrum. Board games, however, are but the niche end of the "revenge" market. Books, book publishers, and bookstores were on economic life-support in the noughties. The emergence of social media and what was called the "attention economy" saw people spending much more time online. Beginning around 2007, the new e-readers such as Kindle, Nook, and Apple's iPad were finding their way into the reading lives of millions. But time spent on digital devices is a zero-sum time. It's time deducted from somewhere else. And time spent on a reading device directly cuts into time spent with paper books (as well as magazines and newspapers). In 2011, Borders bookstore, once the corporate nemesis of the small independent trader, crashed into liquidation—suffering the very fate it had inflicted on many thousands of its monopoly-practice victims before it. However, the bonfire of the book did not burn for long. From around 2014, e-reader sales began to plateau, and book sales began to recover and even to inch up in the United States, the largest market by volume, at an average growth of 0.3 percent over the following five years. Rational and dispassionate sales-figure analysis sees this analog-digital format struggle in terms of an industry "transition" where the future will tell if the e-book has reached its natural level of consumer penetration and will find its lasting place alongside its analog alternate. I think, however, that something more primarily human has occurred in the important "real world" experience of reading a paper-made book: the book has survived because its place and its meaning in our analog heritage is presently too deep to be knocked off its historical course by the digital version of the ancient writing tablet that changed *everything* when it appeared in Mesopotamia about 3,500 years ago. I'll say more about this most ancient and vital analog technology in the next chapter.

Board games and books are only a part of the analog reawakening that Sax and others see as a human response to the alienating experience of pervasive digital technologies. For example, the iPad and iPhone art of someone like David Hockney is a new art form that is widely known and admired. Using apps that simulate pens, pencils, brushes, charcoal, and all kinds of paper, paint textures, and colors, Hockney creates works that he exhibits and sells—albeit as real paper prints and coffee-table books. Thousands of such apps, from basic finger-painting for children to professional standard tools, are now available and sell or are downloaded for free in the millions. However, this development, beginning in the mid-2000s, was paralleled almost synchronously by a booming global art and craft supplies industry. "Real" products such as paints, pencils, pens, and other drawing and craft tools sell more than ever. In 2019 an industry magazine reporting on the good health of the industry wrote of the new craze for DIY art as "expressions of the soul" and linked analog more broadly with the consciousness of authenticity and tactility.

Authenticity has been an analog renaissance keyword that has attached itself in the popular mind to a widening range of cultural practices that connote the rediscovery of another, better, and more authentic way. We see it in practices such as craft beer, freshly brewed (not instant) coffee, repairing things, reusing, recycling, DIY, and much more. The Ilford photographic paper company, for example, which has its roots in Victorian Britain, writes on its website about the growing sales today of film cameras, Super 8 film, photographic paper, and numerous darkroom accoutrements. The blog surmises that such purchases are part of a broader desire to create a sense of "analog community" in the face of inexorable digital virtuality. And

Sax's "real things" link also to the burgeoning "well-being" industry that does much to create the demand for authenticity through discourses of awareness of one's mental health, purpose in life, positivity, and happiness.

Pew Research looked at well-being in the context of analog technology. Complementing a report on the "Positives of Digital Life," they published findings from technologists and educators about digital life's negatives. About washing machines and well-being, for instance, one respondent asserts the refusenik view:

> Did the digital washing machine clean my clothes better? Sometimes. I liked and used about 20% of the options. But overall, I had been perfectly happy with my old analog washer. Was the digital washer more expensive? Yes. Did it break faster? Yes. Was it fixable when it broke? No. Did the digital machine raise my stress level? Yes. Overall, did the digital washer improve my well-being? No. And it wasn't even connected to the Internet of Things, surreptitiously collecting data about my lost socks and water usage. Just because we can make everything digital doesn't mean we should. There are cases where our well-being is better served by simpler, analog tools.[7]

Authenticity was a big thing in Pirsig's motorcycle odyssey. To be closer to nature and to the tools that we have drawn from nature and through which we make and remake ourselves as an integral part of nature has obvious appeal. The world of 1974, when *Zen and the Art of Motorcycle Maintenance* was first published, was very different to our own. But Pirsig's awareness reflects something that began, really, with the nineteenth-century Romantic poets, such as Wordsworth and Goethe, or artists such as William Blake, or environmentalists like Thoreau, who all grasped that a terrible truth lived at the heart of industrial modernity. They perceived an out-of-control machine logic, a genie that was divorcing humankind from the "real things" of the natural world. Today digital automation and virtuality have taken this process to another dimension. The loss of nature that was a feature in the Romantic perspective has been joined more recently by digital-era ideas on what automation and black box complexity means—which is the loss or attenuation of the long accretion of human skills acquired through making things.

Richard Sennett, who has been called "sociologist of the analog world," wrote *The Craftsman* in 2008 and in it he argued that *Homo faber*, or "Man [*sic*] the maker," is disappearing along with the tools and skills that, in shades of Pirsig, were necessary for the making of quality and the qualities inherent in making.[8] Another is the "philosopher mechanic" Matthew Crawford, whose *Shop Class as Soulcraft: An Inquiry into the Value of Work* filled a perceived experiential lack in the minds and bodies of many Americans by selling in vast numbers. Its overseas title was *The Case for Working with Your Hands*. The book was a straightforward *cri de coeur* for a return to the tools, and one reviewer labeled it "a 21st-century update to Pirsig's *Zen*."[9] The clue is in Crawford's overseas book title. We all should get in touch with the actual world, a real and objective world that has dissolved in front of our eyes due to overeducation, robot production lines, office cubicles, and the automation of just about anything. Crawford has a PhD in philosophy, but for a career, he chose to run his own motorcycle repair shop. Every day he works with his hands to engage with the world, to use his own judgment, intuition, and tacit knowledge about how things work and how they might work better. These are skills that we have lost, in the main, through what he sees as the alienating effects of the digital "tools" that have swept away the tactile practices of millennia in just a few decades. And what he sees as the vital human ability to learn and deploy a "manual competence" is, like Sennett's "material consciousness,"[10] being sacrificed on the altar of a post-industrialization that regards such labor as old-fashioned, as "blue-collar," and as belonging to a rusting bygone age that Silicon Valley is upgrading for us all, in the name of progress.

A narrative such as I've just given you, with its analog anecdotes and histories and stories, could go on. There are many more examples that would show a hankering, expressed as a cultural "renaissance," or as an enthusiastic consumer market for analog "retro" products, or as a more diffuse frustration or disappointment with the reality of digital "progress." But we need to pause and ask: Why do many of us feel this way?

It was a question Carol Wilder asked in her essay "Being Analog." Wilder wondered, "What is it about the analog that is so seductive, so persuasive, so 'real'?"[11] She ponders the question at the level of human feeling and understanding and how in modern life this increasingly bumps up against the obstacles of cold calculation and efficiency, even at the seemingly trivial level. And so, for example, she wonders why (in 1997) the "New York subway system owners encountered such resistance when they tried to move riders away from using analog brass tokens to digitized MetroCards."[12] For Wilder this illustrates our ineffable attachment to material things. But we still need to ask: Why is a (admittedly charming) New York City Transit Authority brass token preferable to a throwaway digital card? And why does it even matter?

We can get some insight into why many of us think this way, a way that is often rooted in the past and within inherited ways of thinking—and that is through the living language that both affects and reflects our behavior. And within language itself our largely unconscious use of metaphor gives clue to both the meanings and the human behaviors that flow from these. So, we can ask: Why is it that words that describe analog technologies, often technologies that are no longer in use, still have currency in everyday language? Why for example, do we say that we are *railroaded* into making a decision, or that the impetuous person who does something the wrong way round is putting the *cart before the horse*?

Linguist George Lakoff and philosopher Mark Johnson popularized an understanding of the role of metaphor in how humans communicate their awareness of the world to each other. In their book *Metaphors We Live By*, they wrote that "metaphor is pervasive in everyday life, not just in language but in thought and action. Our ordinary conceptual system, in terms of which we both think and act, is fundamentally metaphorical in nature."[13]

In other words, through our largely subconscious and automatic use of metaphor we convey what is in our heads and in our culture at any given time. Metaphor is embedded in language, of course, but for Lakoff and Johnson it is also "pervasive in everyday life," in our thoughts and actions. This is revealing. Like a secret code written in invisible ink all over everyday language and life we metaphorically *paint a picture* to describe something to someone, just as we *draw a line in the sand* to indicate the limits of this or that situation. And when speaking about the physical person, when we say that someone is *blossoming* or is *wilting*, we know exactly what is being communicated. Metaphors acting as linguistic objects allow us to focus on a concept in an evolutionary fashion that has worked its way so deeply into our unconscious mind that we need little or no training as to its practical operation.

Language is dynamic and the metaphors we adopt reflect an always-changing world. New metaphors arrive unannounced and with no little mystery as to their origin. A way to think about their circulation is as *memes* springing from what the biologist Richard Dawkins called the "soup of human culture."[14] It was Dawkins, indeed, who coined the term "meme," which he derived from the Greek root word *mimeme*, which means "imitated thing"—a word that comes close to the meaning and function of analog. Metaphors are carried through the vector of language. They draw on cultural memory and cultural practice and so may find currency that either persists to embed itself more deeply into language or dissolves and fades away.

Technology-based metaphors are studded through the English language. We see them applied everywhere, such when an editor *takes an axe* to an unfortunate writer's manuscript, or an NFL running back acts as a *bulldozer* to crash through the opposition defense. Some metaphors are ancient and venerable, and politics is replete with them, such as the electoral momentum that *steamrollers* the candidate to victory in the election, or to "pivot," a term from mechanics, but which also means to turn around one's opinion or viewpoint; there is the "snapshot" from photography, but which also means to give the context for an opinion poll. Older metaphors compete with newer ones to become part of the language, such as "moving the *needle*," to indicate a shift in voting intention or political attitude, or to "*dial* down," another metaphor-cum-phrasal-verb of more recent political origin that means to tone down the vehemence of an argument or intensity of opinion.

Metaphor in language tells us that an undead analog zombie dwells in the machine of digital culture and society. It lives not primarily as cultural retro, as in the examples we have looked at, but as cultural memory. But the analog zombie is more than memory—and more than a zombie. It is an evolutionarily formed expression in our language, in our physical form, and in our involvement with the physical world. Analog metaphors denote a *lack* or *void* in our lives, in that we use them to keep the traces and the connections open to our relationship with technology that made us what we are: analog creatures.

Ironically, analog metaphors in our language helped build the conceptual bridge from the solid world of analog to its virtual digital opposite. A part of this cultural-technological process we see in what computer scientist Alan Kay called the "desktop metaphor" that was created to steer the layperson toward a crucial aspect of digitality: the "personal computer" revolution of the 1980s that was popularized by Microsoft and Apple.[15] That the metaphor was a "desktop" was a should-have-been-worrying signal about what the real aim of the personal computer revolution of the 1980s was about. It was about business, and

Metaphor in language tells us that an undead analog zombie dwells in the machine of digital culture and society.

the revolution was a business revolution that swept up the world.

It's a mostly unconscious process, but we're attracted to what are complex machines, designed for primarily business application, through the Graphical User Interface (GUI) and the Natural User Interface (NUI). These rendered the technically difficult navigation around a computer something even children can "understand." The interface is where human meets computer—and where we say goodbye to the analog as the virtual realm says hello. To make computing—formerly associated in the public mind with the military and corporate business applications—feel and look "user friendly," the interface was made to look cartoonish and infantile. And thinking back to *recognition* as a key aspect of what constitutes analog processes, the desktop needed to *have the look* of analog and have equivalences in the real world enough for us to "grasp" the actions we are invited to undertake. The current phone and laptop have a "desktop" that has upward of eighty apps on in it, and there are now as many icon designs as there are icons. But when the public was first introduced to the personal computer, and later the mobile phone, the icon language was simple graphical representations of analog methods or devices. And many remain to remind us of a previous technological world. And so, the clock icon is an analog clock; the mobile phone camera, a camera shutter; the video function, an analog TV; and so on. The WhatsApp icon, a Facebook service used by billions, is an analog telephone handset, a device that many WhatsApp users would never have used, or even physically seen, but would still recognize what it does.

With the app interface we know what we are supposed to do, because it was made easy by making the process look familiar, look culturally appropriate, look analog.

The conceptual bridge from the real to the virtual world was made possible by the icon. Part of the icon's role was to make the transition easy, as just noted, but just as important was to give a virtual likeness of the "seductive," the "persuasive," and the "real" that Carol Wilder saw as the essential attractiveness of analog and is the psychological wellspring of retro culture. Except that they are not. They are icons that represent a technical process that has either become obsolete or has morphed into something else altogether, like the typewriter into the laptop. And so, for example, the old movie film camera icon that is the iPhone's Facetime, or the road map icon for its GPS function, is a depthless metaphor inhabiting a digital liminal space or threshold that opens onto a virtual one-way street. Likewise, the pervasive screen, the general interface, faces in one direction only: into the digital and virtual. But behind us lies a history, a lineage, and a relationship that stretches back for thousands of years.

But we rarely look back, because the Internet is engineered for us to be always looking forward and moving forward in virtual time and space. As the desktop metaphors themselves symbolize and remind us, we can never fully shake off who and what we are: embodied analog that evolved with and through technology, *as* technology, tactile and recognizable technology whose logic is graspable and sometimes reassuring and familiar and *real* in ways that we can hardly explain to ourselves or each other, so deep does the association go. It just feels and looks natural when compared with the virtual. And so, we hanker for the real when we intuit that the virtual doesn't quite cut it in terms of satisfaction and authenticity—which can be often.

And so, the "retro" vogue of analog is only a surface aspect of what is going on. Analog is more than the nostalgia or fun that keeps the revived Crosley Corporation of America and its ilk in business today. Sales of vinyl records and turntable hi-fis could well evaporate, but the not-yet-dead analog zombie in the machine continually calls forth something else back to the culture, like the road atlas in the car's glove box, the manual typewriter (Amazon and Home Depot sell them, handmade in Shanghai), steampunk, photo booths, SLR cameras, fountain pens, CB radio, slide projectors, crystal radio sets, all kinds of electronic audio equipment—choose your own. It's not the actual technology that matters so much as the relationship with it. The feel and the feeling, the perceived authenticity, the lack of precision that corresponds with our own crooked

The White Album and
the LAMY fountain pen
are material things
that connect us to the
material world and
ultimately to nature and
to the cosmos.

timber, the natural attraction to a device or thing or process that we really shouldn't miss if we followed the logic of progress and modernity. But we do. Why would you buy a horribly expensive vinyl copy of the Beatles' *White Album* for $150 when you can stream it whenever you want, for almost nothing? Yet people do. They do because it fulfills something more than an empty consumer urge. They do it because *The White Album* or the LAMY fountain pen and the Schaeffer bottle of ink and the optional piston converters and rubber bladders that we buy even when there is no objective need for them any longer are material things that connect us to the material world and ultimately to nature and to the cosmos of which they and we are a part. And the connection is technology. And the technology is a relationship, a connection, and a story that has a long history.

4

ANCIENT ANALOG: WRITING, COMPUTER, CLOCK

Making Do

The term *making do* has such a positive feel to it. To make and to do. And especially in these days of overconsumption, of oceans of plastic and of throwaway everything, *making do* resonates with a strong ethical tone and a high moral register. Like recycling or the purchase of only local, organic, and seasonal food, it signals awareness of the planet's sustainability threshold and a personal commitment to work within its limits. It has other connotations too. Terms such as "frugal," or "penny-wise" can be summoned up from "making do." For many it can simply mean not having enough money to buy the things we need, and so having to make do by extending the life of something or going without something.

However we look at it, making do as either positive attitude or negative struggle locates all of us squarely in the material world of things and work and consumption and *production*. *Making* as producing. And *doing* as transforming. This has been our way since we evolved to become a technological species. Making and doing is what we evolved to do—and express what we are. And as we saw in the first chapter, our survival depended on it being this way.

Making and doing through rudimentary tool-use enabled us to cling to bare life. In one sense this does not seem very impressive. For most of our 70,000-year history, our knack for making and doing did not stop us from dying young; it was only 30,000 years ago that we began to live beyond thirty years. In another sense, the fact that we managed to cling on at all, "unfinished" and "naturally defenseless" for the trials of life as Arnold Gehlen phrased it, is an astonishing thing. Yet we did. Inventiveness with our environment meant that, uniquely, we were able to colonize almost every part of the planet. We came from the savanna of eastern Africa at a period when the local climate was comfortable, consistent, and Eden-like. However, population increases, intergroup conflicts, food and water scarcities, and other factors compelled some of our species to look for somewhere else to live: somewhere less populated, less hazardous, and less depleted of resources. So, they walked from where they couldn't live any longer to somewhere they could: to new geographies with perhaps different fauna, flora, and climate, and began to *make do* through what humans would become gradually expert at: innovation and adaptation to new challenges and opportunities for our technologically oriented brains and bodies.

Consider this by Helen Epstein on the resourcefulness of the ancient Inuit people:

> The Inuit migrated across the Bering Land Bridge from what is now Siberia and in 1000 AD settled in what is now north-eastern Canada. In the long winter darkness, the wind is so strong that blowing snow can draw blood from exposed skin, and the temperature sometimes plunges to –60° Fahrenheit. In summer, swarms of mosquitoes can exsanguinate a caribou. Nothing grows except berries, moss and wildflowers, so the Inuit hunted seals, fish, birds, polar bears, caribou, walruses and whales. They made houses from snow, skins and moss, and wore fur clothes sewn with sinew threads and needles carved from slivers of walrus bone. They constructed dogsleds from antlers, with frozen fish wrapped in sealskin for runners, and ingenious eye-slit goggles carved from caribou bones that protected them from the blinding light that reflected off the snow.[1]

Such indefatigability resides in our techno-analog DNA and was replicated across the planet over thousands of years and in numberless human stories of disaster,

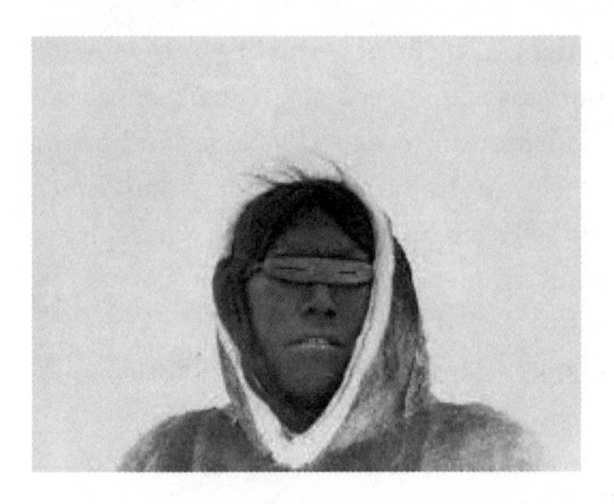

Figure 3 The earliest Inuit snow goggles, dated 1000 BCE

success, or something in between. The Inuit or the Aztec, or the predecessors of indigenous Japanese or Australians, or whomever, would arrive at what for them seemed to be an environment of potential or resonance, and they made do and stayed, or moved on and eventually settled somewhere else, or died out.

"Resonance," as we saw, is that anthropological affinity with one's surroundings, that context of confluence where nature finds its positive echo within our being. We can see this in our own time as a kind of well-being or harmony, a psychological place we can thrive in or a physical environment that has the potential to sustain us. "Potential" in this context means more than its ordinary definition of something *possible*, something abstract, or something random.

In the human-analog-technology connection, potential is something *latent*. Potential is an innate dynamic that is activatable through cognitive problem-solving with new technologies for new environmental challenges. Potential is made active also through an intuition or tacit knowledge that comes from our resonance with technology that made our ancestors both human and techno-analogical.

Arnold Gehlen imputes this dynamic potential into the mind and actions of the earliest technology user. Through intuition or tacit knowledge, the inner voice of an ancestor might say when involved in the daily business of making do, and noticing, for example, the particular shape of a stick, or unusual strength and pliability of a long length of dried-out creeper: "I'll take this along, I might be able to use it." Only humans do this. We do it because being technologically evolved gives us the natural capacity to adapt our surroundings and adapt ourselves *to* our surroundings. Adaptability means also that we walk away when we have depleted the sustainability of a land, or we run away from it in order to flee more powerful rivals, so to begin again somewhere else, carrying our knowledge and our tools with us. Constant movement and change. This gave the potential and latency in the human-technology relationship a *developmental momentum* that is at the very basis of what today we call innovation.

Being analog creatures who resonate with nature and its materials did not mean that the dynamic potential in

“Resonance” is that anthropological affinity with one’s surroundings, that context of confluence where nature finds its positive echo within our being.

the association was promptly unleashed in the lives and labors of our ancestors. For many thousands of years, we struggled to cling to the rock of bare life. Human society was static in ways we could not now easily comprehend. Over the span of 10,000 generations our species changed relatively little or changed only incrementally.

Acquired mastery over fire around 300,000 years ago was a turning point. Fire constituted a replacement technology that gave us an adaptive capacity we did not naturally possess. It was a technique that distanced us still further from other animals in that we were able to transform a landscape much faster through burning and clearing. We were able to cook foods that were previously indigestible, such as wheat and rice, and so slowly learn to cultivate so to become more settled, less dependent on the unpredictable wild bounties of a given area. Settlement was a sociotechnological and *epistemic* transformation that not only implanted us more deeply into our universe by way of understanding the patterns of seasons, of animal migrations, of the various necessary resources of settlement and the new tools required to exploit them. Settlement also meant new stages of socioeconomic complexity, expressed through sustained physical contiguity over long periods of time and through new forms of cooperation, hierarchy, organization, and responsibility. New ways produced further needs, and these spurred the development of perhaps the most consequential analog technology that humans ever produced: writing. By means of the written word, the ancient practice of making do and clinging on was given an immensely powerful cognitive dimension that forever transformed our ways of living in the world, and the physical world itself. The invention of writing, like everything else in our long evolutionary process up until that point, was a *random* event. However, symbolic marks on a medium of stone, slate, papyrus, or paper, once established, set humanity on a path that would become very much *fixed* and would lead directly to the computer age in what in relative terms would be the blink of an eye.

Writing

It's not entirely clear to us how and why it happened, but around the world at around the same time—in Mesoamerica, the Middle East, the Indus Valley, and East Asia—humans began to write. Yuval Noah Harari, in his international best-seller, *Sapiens*, describes the trigger event being populations and property reaching a critical threshold that strained the mental capacity to store all the information needed to keep the society functioning in a more-or-less orderly way: "The first to overcome the problem were the ancient Sumerians who lived in southern Mesopotamia. There a scorching sun beating upon rich muddy plains produced plentiful harvests and prosperous towns. As the number of inhabitants grew, so did the amount of information required to coordinate their affairs. Between the years 3500 BC and 3000 BC, some unknown Sumerian geniuses invented a system for storing and processing information outside their brains."[2]

The earliest examples that we have of that system, or that extension, is cuneiform, abstract representational marks on a medium such as clay tablets. Cuneiform had antecedents in pictographs, or "picture writing," which were ideographic representations of an object in the world, such as a tree, or river, or animal. Writing is therefore an elemental example of the analog human in technological evolution. It is the invention of a tool drawn from nature to represent a social relationship through the communication of an idea. The ideas were first a form of accounting within a community, an aide-mémoire for who owned what and who owed what. Cuneiform was a basic form of literacy, in that a representation of, say, a goat, would stand for the whole word for goat. It was also a basic form of numeracy. In the accounting or recording of who owned or owed goats, the number involved would be represented by a communally understood mark of some kind.

In what was a rapid technological evolution, pictographs evolved into cuneiform, which in turn became scriptural writing. The invention of script was the turning point and accelerated still more the transformation of the human-technological evolution. According to Walter Ong, script were strings of coded symbols, used by "a *conscious* human being as a cue to sounded words."[3] Writing as a technology, for Ong, was not simply an exterior aid but an "interior transformation of consciousness."[4] For him, writing was a "drastic" invention: "It initiated what print and computers [would] only continue, the reduction of dynamic sound to quiescent space, the separation of the word from the living present, where alone spoken words can exist."[5] Stories, lessons, techniques, warnings, behaviors, skills, and other crucial information were no longer just orally transmitted and so necessarily inexact through being passed from speaker to hearer time and again over stretching time and space. The extent of this psychological transformation in humans and their societies is difficult to put into words. We get a sense of it when one tries to imagine an alternative path that we might have taken without writing; it is almost impossible to visualize.

The word and the idea that it represented became abstracted from our "living present" and was fixed in lasting time and space upon media such as stone, clay, or papyrus. Writing was then a record, a trace, and a "proof" or confirmation of events in the lives of humans. With a written record of an event such as a transaction, an obligation, or the rendering in writing of a human *story*, human society became more than the disparate and unconnectable actions between multitudes of people in a real-time physical present. Human relationships now had a record. There was now

Figure 4 (Top left) Sumerian pictographs, the earliest papyrus date to around 3100 BCE; (top right) clay cuneiform tablet, unknown creator, c. 530–522 BCE, under Cambyses II of Babylon, British Museum; (bottom) letter written in Greek on papyrus from 310 BCE

a recorded past that was real because it was consequential, and it was consequential because it was recorded. Humans also had a future—an agreed and recorded commitment to do something at some point forward from the time of recording. Humans passed from the time "before recorded history" to the time of history and became part of history because it would be recorded. Writing therefore instilled upon human consciousness a perception for the passage of time, and as we will see later, this acquired a technologically enhanced facility that would directly enable the invention of the clock, another major analog technology that has had inestimable effects on our species and on the world in which we live.

All this is an immense burden of significance to place onto the technology of writing. But it is justified. Walter Ong reiterates the point in an essay titled "Writing Is a Technology That Restructures Thought." With just a hint of drama, the American Jesuit scholar argued that writing "takes possession of [human] consciousness." He goes on to claim that script is endowed with almost supernatural powers in that "it tends to arrogate to itself supreme power by taking itself as normative for human expression and thought."[6] And so, in a very real sense humans became the *product* of writing, of literacy and numeracy. They were also subject to the effects of a technology that has "emerged" from them—and then abstracted *away* from them and their analog natures. These technologies transformed us from being creatures possessed of limited stores of oral and tacit knowledge and constrained powers of cognition to becoming a super-species with access to ever-expanding realms of limitless sources of abstract knowledge to create and draw on. By transforming ourselves thus, we were able to change the physical world from a point of omniscience that is denied all other species.

Writing served to codify knowledge and laid the basis for the rational organization of humans, societies, cities, civilizations, and cultures. In the West, the long prehistoric oral tradition with its epics and legends containing heroes, gods, wars, and poets was doomed to be overrun by a new mode of communication. Around 370 BCE in *The Phaedrus*, Plato denounced the written word as quasi-knowledge of a particularly rigid and memory-destroying sort. Nonetheless, it was ironic that it would be writing on which Plato's philosophy—and all the footnotes to him—would be based. Through the written Greek, its philosophy became the foundation of the classical arts and sciences in the West, and these disciplines, the creative and the systematic, are the pillars upon which Western and later all modern civilizations were built.

Computer

Despite his suspicions about the utility of writing, Plato used it as the very basis of his Academy, the very first recognizable university, which he founded in 387 BCE. In it, knowledge was harnessed, concentrated, and expanded through writing. Dialectics, philosophy, and political theory, for example, were discussed and taught alongside mathematics. The latter seems to have had special status in the Academy, as it is said that at the entrance to it there hung a sign that read: "*Μηδείς άγεωμέτρητος είσίτω μον τήν στέγην*," which translates as: "Let no one ignorant of geometry enter here." Geometry is concerned with the properties of space, and this is significant for our story. Plato instructed his academicians to use geometry to discover how the recurrent and ordered movements of the stars and the planets could be understood. And it was their "solution" to the astronomic riddle of the constantly changing night skies that led to the design and construction of the first computer.

So it was that only 200 or so years after the founding of the Academy, advances in geometry and practical mathematics produced at least one dazzling result. And it is one that as far as analog technology is concerned, plots a direct line from that time to our own. Around 100 BCE, the so-called Antikythera mechanism, a hand-cranked, multigear, wheeled analog computer, the very first machine computer, was built based on the findings of Greek astronomers and mathematicians.

The Antikythera mechanism was discovered in 1901 aboard the sunken wreck of an ancient Roman cargo ship. It was a bronze and wood lump, badly encrusted

Figure 5 (Top) Fragments of the Antikythera mechanism; (bottom) envisioned replica

and deeply corroded, and would have been about the size of a shoebox. The device was built to calculate and represent celestial movements, especially lunar cycles and the luni-solar calendar. It was also thought to have once mechanically displayed planetary movements. However, this highly advanced mechanism was more than an interesting conjunction of engineering and astronomical prowess for the appreciation of the ancient Greek elites: it was a fundamental expression of the survival instincts of our technologically evolved makeup. Writing in the *Nature* in 2006 a group of experts on the mechanism reasoned: "Eclipses and planetary motions were often interpreted as omens, while the calm regularity of the astronomical cycles must have been philosophically attractive in an uncertain and violent world."[7] In other words, this very first computer was a form of psychological and technological *control* for humans, control over their social and cultural world, and control over their known universe. Bringing order to a capricious world through science and philosophy reinforced the Aristotelian concept that humans stood apart from nature and that their technologies were analog of it—but that somehow, they, as human, were not.

The fabulous ideas underpinning the Antikythera mechanism were lost to the world for over a thousand years with the collapse of Hellenistic Greece and the fall of the Roman Empire. What this bronze and wood minicomputer represented, however, was the rapid and extraordinary advances in the arts and sciences made possible by writing. In its original glory it would have triumphantly signaled that materially and intellectually, humans no longer clung so tenuously to barest life, but were now masters of their destinies. The revolution of the mind that was writing was also a revolution in how humans acted in the world. Through analog technologies that allowed people to express their own analog essence, humans were able to impose order on their world and regulate their societies to be less random and threatening to life and culture. As European societies began to emerge from the Middle Ages to rediscover their Greek and Roman heritage, the rhythms of the planets and the seasons that the Antikythera mechanism had plotted through its intricate gears and inscribed dials began to assert themselves as a new appreciation of time, and assert themselves also in a much more granular way on the everyday lives of people. Most importantly, through the writings of Aristotle time had come to be understood as a "series of now points," as something calculable in number, as something that a machine derived from the Antikythera mechanism could materially represent. Such a machine emerged in the revolutionized form of the mechanical clock.

Time and the clock, like writing and reading before it, initiated another profound change in the human condition. With the analog clock as a material representation of time, order and control and rationality were able to escape

Through analog technologies, humans were able to impose order on their world and regulate their societies to be less random and threatening to life and culture.

the interior mental abstractions of the philosophical elites of the Classical era and become a physical presence in the life of the individual who, as a matter of necessity, would need to learn to "tell" the time so to synchronize their life and experience with its mathematical meter.

Clock

If proof were needed that we do not stand apart from nature but are analog with it through technology, then we only need consider the biological clock that exists in us as well as in plants and other animals. Our in-built clock, residing at the molecular level, operates on a twenty-four-hour cycle and functions so to give rhythm to our sleeping, eating, and wakefulness patterns. Our human clock is circadian (from *circa* meaning "around" and *dies* meaning "pattern") and was formed in evolutionary adaptation to the change from light to dark in the environment. Taking our cue from the environment was therefore an even more deep-seated trait than Democritus or Aristotle realized when they considered our relationship to technology and to the natural world. Current chronobiological research on the circadian rhythm stresses its vital importance for both physical and mental health. We are genetically predisposed to be either "larks" or "owls" in our sleep-wake rhythms. However, to be out of sync with the rising and setting sun, such as when working the night shift for prolonged periods of time, can be very bad for your health. The World Health Organization, for example, has said that extended periods of night-shift work may even have a carcinogenic effect.

Well before we grasped the modern science of our relationship to the movements of the planets, we knew that we had an important resonance with the skies above. The movement of the planets was always full of meaning. Aztec and Egyptian civilizations worshipped the sun, and lunar, stellar, and other deities existed in many cultures of the premodern world. Importantly, there was diversity to the human relationship to the cosmos that reflected the different understandings of what ancient people interpreted from the firmaments and their changeable patterns. Differing perspectives on the constellations, and the short and long cycles of the return to view of certain stars or planets, gave the universe diverse local significances. Chinese, Islamic, and Christian calendars, for instance, all had differing reference to the cosmos as interpreted by the local scholars of the time who created them.

The imperfect analog of the heavens matched the crooked timber that is humanity. The Antikythera mechanism, however, was a product of reason, writing, and mathematics. It plotted the known universe and represented it in a way (and through a machine) that was unerring, unchanging, and exact, with its gears producing the same revolutions every time. The clock, with its mechanical representation of the twenty-four-hour cycle, emerged straight from the logic and philosophy of the Greek computer. And like the philosophy that underpins the Antikythera machine, the clock was adopted as a form of time reckoning that was oriented toward the imposition of a universal order and social control. Through mathematics, the crooked timber of humanity would be straightened out, or so it was thought, by the modern advocates of clock time.

The earliest mechanical clocks date from early fifteenth-century Europe and were used by ecclesiastical authorities to signal when bells should be rung for prayer, or to call the peasantry to Christian worship. By the seventeenth century, and through innovations in science by Isaac Newton, in particular, the mechanical clock and the form of time it represented came to be understood as a fundamental scientific truth about what time is. Newton's celebrated *Principia mathematica* of 1687 formed the basis of classical mechanics and the laws of planetary movement. Strengthened by his reputation as one of the most brilliant minds of early modernity, Newton argued that the universe is like a vast clockwork mechanism and that the movements of the planets and stars is literally the movement of time itself. As he put it: "Absolute, true, and mathematical time, of itself, and from its own nature, flows equably without relation to anything external."[8]

Following in the Greek tradition, time, for Newton, was a process from which humans were separate, a kind of "container" within which humans existed. Absolute time in this view was God's creation, with Newton dutifully supplying what N. W. Chittenden, in his "Introduction" to the 1848 edition of *Principia*, called the "master-key of the universe."

Newton wrote a good deal about the clock in *Principia*, arguing that as a mechanism it needed to acquire ever-greater accuracy through a "rational mechanics" so to synchronize with the God-made universe. Consequently, humans, in their growing synchrony with an always more accurate time mechanism, would bring themselves closer to God. From the eighteenth century onward, punctuality was seen in the middle and upper classes as the mark of dependability and an enlightened approach, and clocks and pocket watches themselves were ranked as signs of distinction and wealth. The ubiquity of timepieces burgeoned rapidly and public places such as coffeehouses and taverns could afford to display them for patrons who needed to "know" the time.

The convergence of what has been called "clock time consciousness" with the rise of the Industrial Revolution was not coincidental. Clock time would form the temporal basis for an industrial way of life coordinated and rationalized by a pocket-sized machine. To its rhythm the new industrial classes were drawn into a new form

Figure 6 Painting of the interior of a tavern, with prominent wall clock, Adriaan de Lelie (1755–1820)

of *organization* through abstracted time. The rhythms of the seasons, of diverse ancient customs, and what social historian E. P. Thompson called the "task oriented" ways of marking and experiencing time were replaced by the clock-timed "working day," by payment by the hour, and by the conversion of time into a tradeable commodity that could be bought, sold, saved, wasted, or lost.[9] Benjamin Franklin famously stated that "time is money." This was a valuation that anyone who worked in industry, as either buyer or seller, was already acutely aware of. The clock with its strict division of the day into seconds, minutes, and hours represented the quantification of human time experience that flattened out all its ancient and crooked qualitative and subjective diversity. With industrialization the social world began to acquire new temporal rhythms that were set around the machine, and clock rhythms of industry became quickly normalized as part of modern life. So quickly, indeed, that as the philosopher Aran Gare put it: "the permeation and domination of life by abstract clock time [was] so complete that it is difficult to realize [today] just how extraordinary this is."[10]

Yet it is extraordinary. Just as we take writing as "normative for human expression and thought," as Walter Ong argued, and just as we assume that science can express the reality of the universe through number, as the Greek designers of the Antikythera mechanism assumed, then the clock and the form of time it represented came to mean time *itself*, notwithstanding the *diversity* of time experience that had existed across human cultures for thousands of years. This subjectivity was something expressed as a conundrum by St. Augustine of Hippo, some fifteen hundred years ago, who famously wrote in his *Confessions*: "What, then, is time? If no one ask of me I know; if I wish to explain to him who asks, I know not." Augustine suggested that time as experience is "in" us in the form of experience, and not "out" there in the form of an abstract concept that is measured by the movement of the planets and represented by the gearwheels of an ancient computer and early modern clock.

Writing, computer, and clock are ancient analog technologies of the first importance. All represented a fundamental cognitive transformation of our species vis-à-vis the relationship with technology and the capacity to transform our environment. But it was a path of increasing abstraction in thought through writing, through the acceptance of the "reality" of the world as expressed by numbers and through computers, and it was a form of colonization by a universalized conception of time that displaced or suppressed the diverse and subjective relationships with time that cultures all over the world had lived with for thousands of years.

Singularly and in combination these analog technologies humanity was set on a path of order and rationality whose metaphors of science and technology were

We are seeing this loss of the "real" and of the human-within-nature much more clearly today in our digitally connected virtual world.

expressed by the machine—a technology itself set on a path of development and sophistication that would one day see machines that can read and write, machines that can calculate unimaginable numbers, and machines that compress time and space by calculation and communication at incredible speeds. Following a rationalized path of technology development, writing and the clock represented stages in the gradual "loss" that Bill McKibben lamented at the beginning of this book. These were world-historical developments, to be sure, but with their abstractive qualities they represented also forms of "disconnect" with our technologically evolved species being. We are seeing this loss of the "real" and of the human-within-nature much more clearly today in our digitally connected virtual world. But there were still steps to be taken down that path. The modern period of analog machine technology was the unavoidable route to where we are today, and so it is to this that we now turn.

5

MECHANICAL ANALOG: CONQUEST BY THE MACHINE

With the arrival of modern machines, our analog essence and the technological relationship on which it was founded began to change. And the change was profound: world-shattering, even, such that humans with their ancient and diverse cultures and forms of knowledge, and without forethought or plan or purpose, embarked on what we can see in retrospect was a narrowing path dependence that would lead, with implacable technological and economic momentum, to digital computers and all that they have meant for our world today.

In this chapter, three case studies—the printing press, the Jacquard loom, and the typewriter—will guide us through a remarkable analog journey in which humans would change from being symbiotic though not particularly efficient inventors and users of tools who would "get by" more or less for many thousands of years in a context of low-impact survival to becoming hyper-efficient, high-impact marauders of the planet by means of machines from which we increasingly become disconnected and alienated, and which jeopardize the nature of our survivability and compromise the ecological equilibria on which our survival has depended.

Tools became machines because they were invariably planned in a short-term way to replace some or all of our cognitive and physical capacities in this or that function. In some ways this made them a rival of the worker. But machines were alternatively seen as complementary to our cognitive and physical capacities and as instruments of progress that established our self-conception as masters of our destinies. Much of the history of technology has therefore been about either the destruction of one way of life or the positive enhancement of it. And it was within this philosophical-technological tension that our analog substance and our evolutionary relationship with technology would transform.

Man Machine

René Descartes, one of the most illustrious mathematicians and geometrists of early-modern Europe—then a ferment of the emergent sciences—said something startling then about a seemingly different subject altogether,

Much of the history of technology has been about either the destruction of one way of life or the positive enhancement of it.

the human body. Writing in his 1664 *Treatise on Man* he declared that "the functions which I have attributed to this machine (the body), as the digestion of food, the pulsation of the heart and of the arteries . . . *imitate as nearly as possible those of a real man*; you should consider that these functions in the machine naturally proceed from the mere arrangement of its organs, neither more nor less than do the movements of a clock, or other automaton, from that of its weights and its wheels."[1]

In contemporary parlance this translates as the body is *just like a machine* in its workings. Humans are a biomechanical mass of interconnecting parts that in its functioning finds its analog in a clock or some other intricate machine. Less intricate was his likening of the heart as a kind of furnace that heats the body and pumps blood around it like a system of plumbing. For the Frenchman Descartes, as for many writers of the time who found something revealed in the universe through science, the body and its wonderful complexity was the creation of the hand of God. Our similarity with machines, moreover, was of a piece with a larger similarity with the wider universe. Descartes saw the cosmos as a series of gigantic vortices that swirled around in an immense mechanical dance, as created by God at the beginning of time. If this "mechanical philosophy" sounds familiar, it's because it's the stuff that Isaac Newton imbibed from Descartes' writing when he was a student at Oxford.

It was Newton who said famously, "If I have seen further, it is by standing on the shoulders of Giants"—which did not necessarily mean that he agreed with them. One difference with Descartes on the nature of the mechanical universe and its links with humans is that while both saw the clockwork cosmos as God's work, Newton felt that his revelations were a providential sign that humans, as fallen per the Old Testament teaching, would be doing God's work if they synchronized—that is, sought perfection—in their cultures and societies with the handiwork of the "most perfect mechanic of them all." Descartes, on the other hand, saw humans more meatily as *bodily imitations* of the mechanical universe or, as I would put it here, as *analogs* of the cosmos *and of their machines* as opposed to standing apart from it, as Aristotle (on whose shoulders they both stood) claimed. What both Newton and Descartes did, however, was to elevate early modern science and technology to a higher level of sophistication in that they saw humans, the cosmos, and technology as being much more closely intertwined, and sharing a common logic with a common creator.

In the mid- to late seventeenth century, Europe was ablaze with generalized intellectual energy. In addition to Newton and Descartes, there was also Francis Bacon, whose revolutionary ideas on scientific method provided a major impetus for the technological element of the coming Age of Enlightenment. And there was Gottfried Leibniz, whom we'll come to in a minute. These were just some of the preeminent giants of early modernity. Such individuals were immensely influential not only in their work on the questions of the time but in defining what those questions were. Their research agendas helped shape Western thinking on questions of science and technology, and their legacies not only produced the material world in which we live but also ensured that it would be science and technology that would direct the route of human "progress." Leibniz, for example, through his efforts with the practical applications of calculus, laid the groundwork for modern computing with the development of the binary system of numbers that is at the heart of computer software today. But much more was happening, and much of it at the level of the practical and concrete, and in the lives of ordinary people. Incremental mini-revolutions were occurring all over Europe, in textile manufacture, in weaponry, in metallurgy, in agriculture and water supply, in building techniques, in tools of every kind and application, and also in the social organization of work. On the face of it, across the countries of early modern Europe, all this appears as the disparate efforts of solution-seeking individuals, guilds, and associations that thought and acted locally, responding to local conditions and challenges. Yet these were all intimately connected.

What was the source of this intellectual ferment? What was it that caused this reshaping, almost literally, of the human mind and how it thinks about the world? Stressing unwittingly, perhaps, the concept of analogicity, Bacon had called his time "the happy match between the mind of man and the nature of things." There was, of course, a mediating force that brought together this "happy match." This was, in a term that sounds so modern and so digital today: media and communication.

The Gutenberg Universe

Measured by today's standards the fundamental source of all this change was a relatively slow burner. But in due course Gutenberg's mechanical printer would change just about everything. The printing press has been said to have technologized the word (although the word was already a technology, as we have seen; it just became much more widely available). Nevertheless, the mechanical printer would also become—and this was a dream of Bacon's—the basis for a *technologizing of knowledge*, to produce a new human mind, a science- and technology-driven mind, which, as he saw it, would be a mind "constantly controlled . . . by machines."[2]

Developed in 1439 by Johannes Gutenberg, the movable-type printing press was the analog technology that not only technologized knowledge but industrialized it too. It made possible the conception, realization, and implementation of the kinds of machines that would become the basis for the Industrial Revolution. The products of the printing press, such as books, pamphlets, newspapers, and journals, became the first mass-produced commodities that would circulate the ideas that would create and shape modernity, and these commodities were enthusiastically consumed by a ready market that was limited only by the extent of its literacy.

Figure 7 The Gutenberg movable-type printing press

This new commodity of print was a contagious push technology whose very presence encouraged its use, inciting people to learn to read or to improve their existing proficiency. The more printed matter that circulated, the cheaper it became; and the more integrated print became within culture and society, the more that literacy was seen to be a valuable and tradable technological skill. Movable type was revolutionary in other ways too. First, it enabled exactitude of copy, with each printed run being identical. This meant that the media afforded accurate communication on a mass scale, something not possible before. Every reader saw the same words from pages replicated by the machine instead of the error-prone and slow hand of the human copyist. Such precision was, of course, a boon to the transmission of scientific ideas and practices. Experiments could be described somewhere and then replicated somewhere else; formulas could be exactly quantified; calculations correctly shown. The inevitable mistakes of the copyist, the predictable errors of the note-taker, and the unavoidable inaccuracies in the retelling of an oral communication were drastically reduced in this first information society. The Gutenberg machine produced for the first time in history the basis for coordination in thought in science and technology, and on a rapidly widening and deepening scale. Second, and related to the last point, the productivity of movable type was phenomenal. It has been estimated that 160 machine printed pages could be produced in the time it took a skilled copyist to complete just one—an increase of 16,000 percent. On a practical level this meant that between the first printing of the Gutenberg Bible in the 1450s until the end of that century, some 200 million copies of these expensive publications were produced and distributed throughout Europe. Much cheaper prints of work in areas such as science, philosophy, and politics could be made at an even faster rate of productivity and made even more widely available.

It was through such possibilities that a new world based on the rigor and exactitude of science and numbers could begin to be constructed. Analog print media may have exponentially expanded Bible readership and with vastly increased numbers of people now able to read it in their own language as opposed to Latin, but more importantly it also promoted an intellectual and philosophical awakening in Western culture. Immanuel Kant, for example, answered his own rhetorical question "What is Enlightenment?" when he wrote in the revolutionary decade of the 1780s that it is "freedom from self-imposed tutelage" and the courage to "dare to know." Kant saw the intellect and the physical body as becoming free through knowledge and through advances in science, technology, and philosophy. The Enlightenment was therefore the possibility of freedom through a new media and communication technology that spread the word, a different word from what had originated from the "tutelage" of the Bible for the previous thousand years. For Kant and many of the *philosophes* of the time, what would be termed "print culture" brought reason to the people and offered the chance of what was seen to be the full flowering of human potential.

The Enlightenment would run its course and its legacies are hotly debated today. However, the products of Gutenberg's machine and its antecedents had an undoubted effect on the thinking and acting human. Through mass-printed writing and knowledge, literate people would become a living expression of the printed word. Moreover, as actors in the world they would change that world in conjunction with the knowledge that print would provide for them and which they would add to through their own writing. In other words, although the medium of print was very much the message, it was not the sole agent. The analog resonance or ongoing interaction between print and human unleashed human potential along a particular path of development. Print culture took many different forms in many distinct cultures. However, its general thrust through the forms of knowledge it produced was the growing sense of human "progress," the awareness among the elites, at least, of a "modern" age that was underway, one that had taken the best of Classical Greece and the Renaissance, and where humanity now not only was "daring to know" through science, technology, and philosophy but was also *actuating* it in the remaking of the world. The

Gutenberg machine was thus a generative machine that made possible every machine that came after it, in that the massively expanded production, circulation, and consumption of knowledge gave humans the capacity to think in more complex and machine-oriented ways. As McLuhan argued, through the printing press, modern humans acquired Gutenberg minds.

Machines need sources of power such as muscle, wind, steam, or electricity. But to be dynamic in their development as better, faster, cheaper machines, they needed something additional to stop from stagnating, and to stop society stagnating with it. Modernity was classically described by Marx as an environment where "All that is solid melts into air"—a compulsory modern context of unceasing and rapid change. But mass print culture and all the books and scientific journals in the world would amount to not very much had they not been infused to at least some degree by the most revolutionary form of dynamism within modernity: capitalism. To understand how this energy works, we need to consider our next essential analog technology and its place within the Industrial Revolution.

Wooden Software: The Jacquard Loom

Software, in modern parlance, is a set of instructions that allows a process to begin and continue along a preset path with minimal or no human involvement. This is also automation. And automation automatically places a space, a breach, between human and machine. When the human gets in the way of a large-scale production process by way of fatigue, capacity, mistake, or the capacity for independent decision-making, that is, refusing to work, then employers have always looked for ways to automate. And the ability to program a machine, to give over and expand roles to it at the expense of the worker, constituted a major step away from analog and toward digital.

A familiar high school history subject, at least in the Anglosphere, is the Industrial Revolution. The importance attached to it is justified, given this is from where it emerged, took off, and sustained its most powerful dynamism for over two centuries. Common themes students are taught are such things as the terrible conditions of the working classes in the forced transition to factories; the enclosure movement in Britain where people were evicted from land to make way for sheep; the slave-powered cotton industries in the United States and Caribbean that supplied the textile mills; the mills themselves, given poetic expression as "dark satanic mills" by William Blake in 1811, and then given sociological notoriety by Friedrich Engels in 1845, with his excoriating *The Condition of the Working Class in England*; the destruction of the Indian cotton industry by the British flooding their markets with cheap machine-made products; the horror stories of child

The ability to program a machine, to give over and expand roles to it at the expense of the worker, constituted a major step away from analog and toward digital.

labor; and the invention of the steam engine that helped supercharge the productivity of machine-made cotton.

Note that there is a lot about mills, cotton, labor, and machines in many Industrial Revolution narratives. That's simply because cotton manufacture was the Industrial Revolution's initial and key propelling force. The production of cotton as a commodity was the economics, the politics, the geography, and the sociology of modern industrialization. Technology was its end-point driver. In ancient Greece, there was no competition that we know of to upgrade or improve or make obsolete the Antikythera mechanism. This is because the objective of its makers was to demonstrate the advances in human knowledge about the world through conceptual innovation and technical skill. However, competition is the very basis of capitalism, and capitalist competition is concerned only about ways to make profit. In that sense, capitalism was both cause and effect of the Industrial Revolution. For the individual textile manufacturer, success in finding technical ways to produce cotton cheaper and more efficiently ensured industry leadership and vast profits—until, that is, another competitor devised something faster and with less input cost. For example, the water-powered mill was the industry standard until James Watt's steam engine of 1776 became a game changer, enabling not only mobility insofar as where production could take place, but also an exponential increase in how much power a machine could deploy.

In its turn, electricity-powered machines would soon overtake the dominance of steam. So it was with another technology, developed in 1804, and which transformed the cotton manufacturing end of early industrialization: the Jacquard loom.

The hand-weaving of cloth goes back to at least 6000 BCE. The thing about hand-weaving is that it was done by hand, with ancillary tools such as shuttles, sticks, and bars all requiring the manual dexterity of the loom operator. Speed was limited mostly by the energy and skill of the worker. This was not a major problem until much closer to our own time. Ancient communities and more recent peasant communities would weave cloth for local use or would barter and trade in markets or along trade routes. Quality and design were the prized value-added feature in precapitalist societies. Change in craft-weaving was slow, but change had been slow everywhere for thousands of years.

Fast-forward to the age of industry. Joseph Marie Jacquard was a wealthy eighteenth-century weaver and inventor. But given the times, he was also a businessman, someone in the business of profit, but also someone unavoidably in the business of competition and the concomitant never-ending search for the competitive edge. He found it in his eponymous loom. Jacquard's machine was revolutionary in one main respect that makes it an essential analog technology: it was programmable.

Figure 8 A Jacquard loom

Drawing explicitly from Gottfried Leibniz's work on the binary system published in 1703, Jacquard applied its logic to the loom to produce the most important manufacturing machine of the time. Binary code is a series of steps where decisions can be programmed into a process through a yes-no/on-off/proceed-stop instruction set.

Installing the code into the machine through a system of punch cards that represented steps in the patterning in a weave, the loom effectively had a stored knowledge that, depending on the punch-card program applied, could produce patterns that were as simple or as complex and aesthetically attractive as the designer, or coder, could make them.

But there was more to the invention than pleasing patterns ingeniously generated by wooden software. First, Jacquard's loom brought a central prototype of analog technology into the logic of modern machines and, by extension, into modernity and industrialization. The punch-card system has a direct connection with the technology of writing in that it was information, knowledge, abstracted from the heads of people and stored in time and space as a fixed and endlessly replicable extension. In developing his binary language, Leibniz was clear what he wanted to do in this respect. It was to create a universal language, or *characteristica universalis*, a formal language that scientists, philosophers, and mathematicians could all share in their discussions, and, as Leibniz rationalized it, any "errors, except those of fact, would be mere mistakes in calculation."[3] The Leibnizian ambition, therefore, was to create for the entire world a perfect, mathematically precise language that would be free of miscommunication, of mistakes in logic or reasoning, and that would advance science and technology for the benefit all peoples, nations, and cultures. Such a glorious vision, as we will see in a later chapter, is the very same one (and from the same technological impulse) that drove the cyber pioneers of the 1960s in the United States and beyond who would go on to construct the Internet.

Second, the programmable loom was able to do something unprecedented up to that point: produce cloth of exactly the same pattern, time after time. Mistakes were taken, literally, out of the hands of weavers by a loom that ran on a program that could not make any deviation to its binary instructions. Consistency was a major value-added quality in Jacquard-made cloth. Of no less importance (from the perspective of the mill owner) was the eradication of worker fatigue—a contributing factor to mistakes, and hence wastage, in the production process. Jacquard's loom began to appear in Britain in the early 1800s, just as steam power began to flow through industries of every kind. The historian E. P. Thompson wrote of the "Jacquard principle," which, combined with steam, pitched workers into "a losing battle against the power of steam."[4] Part of that battle was expressed through the Luddite movement. These were groups of workers who, following the lead of a perhaps mythical character called Ned Ludd, took to smashing the power looms that were seen by hand weavers to be job-destroyers. The importance of the Jacquard loom for the owners of cloth mills, a business then at the spear-tip of the Industrial Revolution, and so of capitalism more broadly, was reflected in the passing of the Frame-Breaking Act of 1812, which upgraded state retribution for the wrecking of looms from penal transportation to the colonies to capital crime.

Third, steam power and the Jacquard principle combined to create, in a rudimentary fashion, the realization of another, more ancient principle, or pipe dream: *automation*. The future reality of full-fledged automation would signal the most complete "disconnect" of the human from machine technology and the diminishment of the cognitive and physical ability to act on the social and natural world through such ability. Many ancient cultures had conceived of automata, mechanical beings that could move and act like people. The lack of adequate technological agency meant that these remained dreams and ambitions that could never be made material. However, armed with the advantages of the scientific and industrial revolutions, philosophers and inventors looked back once more to the Greeks for their inspiration in this regard. They looked to Aristotle, for example, who in *Politics* speculated that if a machine "could accomplish its own work [and] the shuttle would weave," then humans would be unburdened by the baleful demands of manual labor; he continued that "if every tool, when summoned, or even of its own accord, could do the work that befits it, . . . if the weavers' shuttles were to weave of themselves, then there would be no need either of apprentices for the master workers, or of slaves for the Lords."[5] It would be a very different world, in other words, one that had been turned upside down.

The Greeks never did replace their slaves, and the dream remained just that. However, the enlightened moderns of the late eighteenth century were able to make progress of a sort. In a fad that swept Europe, ingenious miniature and life-sized clockwork automata were constructed: human-like dolls that could "play" musical instruments without mistakes, such as David Roentgen's famous 1784 creation of "Marie Antoinette" playing the dulcimer. Such playthings would enchant the public—and would enchant the actual Marie Antoinette when she was presented with Roentgen's imitation of herself. A couple of decades earlier Jacques de Vaucanson constructed the "Flute Player," a life-sized automaton complete with a sophisticated system of bellows to supply "breath" for the flute. This deeply impressed the French Académie des Sciences and so they gave him a medal for his trouble. Indeed, it was Vaucanson who pioneered the punch-card system for a loom, which deeply impressed Jacquard, and who promptly used it for his loom.

This was progress of a sort in that such automata showed how far we had come in such a relatively short space of time. For our purposes, though, the desire of humans to replicate themselves as automata, a desire that crosses civilizations and cultures, can be seen to be a very *analog* desire. Automata are extensions of our bodies in that the body (if not the mind) is completely replicated to exist in space and time as an abstraction of ourselves, but it is also an *imitation* of our bodies, an analog of humans, just as the airplane is both imitation and analog of the bird. Automata therefore express something deep within us, something perhaps that motivated Descartes in his thinking that the body is analogous to the machine and therefore that the most perfect technology possible is the ultimate analog of ourselves—a thinking and moving robot.

Last, and importantly, there is a regression of a sort here too. And it's a regression of a more readily understandable kind in our own time. Machine automation—having the machine do the work, in whole or in part—nullifies, in whole or in part, the active part of the human in the ancient resonant interaction with technologies that has characterized the relationship since the very beginnings of our species. The Industrial Revolution industrialized this degradation of the labor process. For Marx, the machine was the very driver of capitalism. The more sophisticated, powerful, and productive the machine became, he argued, the more the worker suffered from unemployment and deskilling. With every new advance in automated production, skilled workers would find themselves as semi-skilled, and the semi-skilled, in turn, are rendered as unskilled or unemployed.

In his "Fragment on Machines" from *The Grundrisse* of 1857, Marx condemns what the factory machine does to the waged worker: "No longer does the worker insert a modified natural thing as middle link between the object and himself; rather, he inserts the process of nature, transformed into an industrial process, as a means between himself and inorganic nature. . . . He steps to the side of the production process instead of being its chief actor. In this transformation, it is neither the direct human labor he himself performs, nor the time during which he works."[6]

In plainer language, Marx becomes the first to identify a quintessentially modern malady, something that would have a profound influence on systems of thought ranging from sociology and psychiatry to economics and politics: *alienation*. Automation, even in its rudimentary nineteenth-century form, did something not exactly unprecedented in the long human connection with tools and production. As with writing and print, it weakened the links and therefore the originating resonances between human, technology, and nature. And by stepping to the side of the production process, the worker under capitalism becomes formally detached from the concept of *Homo faber*, of the human as maker, as a participant in the interaction with technology and nature.

It is common to speak of alienation today in reference to a whole range of social problems, from drug use and unemployment, and from mass housing projects to Internet addiction. However, we can see these as surface manifestations of something deeper in our relationship with tools and of our analog nature. Automation is never total. It still takes humans to build robots. Workers still need to mind and service automated processes. And automation creates new general service jobs that did not exist before the latest automation "solution" in this or that industry. What the Jacquard loom did, though, was to *begin the process*, to show proof of concept that machines could do what the ancients dreamed of. And we still have those same automation aspirations—the dream to be free to do other things, to be part of the leisure society that was anticipated at the beginning of both the industrial and digital revolutions, to benefit from the social dividend we were told would accrue if we allow machines to do more of the work for us.

Yet it has never worked out that way. Industrialized societies are today blighted by a double alienation of joblessness and underemployment for many millions, and by those in the overworked stratum, numbering also in the many millions, who have little time for anything other than a work process that blurs the boundaries between work and leisure, home, and workplace. It is therefore something of an irony that in the quest to imitate our analog selves to make work easier, if not redundant, the process of automation has qualitatively shifted our relationship with technology. It has become in our own time a strange relation, one characterized by a distancing and lack of intimacy with the machines that make our world (for us), a relationship that the philosopher Rahel Jaeggi called—in a term that captures something of automation's spiritual emptiness—a "relation of relationlessness."[7]

Productivity is the central and elemental core of the logic of automation. Without the need to do things to faster to produce more cheaply, there would be no machines as we know them today, no PCs to help us write faster and no Internet to drive the information revolution, which, after all, is merely the expression of the ever-faster processing of data by silicon chips. Writing by hand was never troubled too much by the need for speed. The productivity of the medieval scriptorium was fast enough for a Church hierarchy that no competition in a largely prescientific, preliterate and precapitalist world. That would change when science, literacy, and capitalism were in the driver's seat of social change. A good example of this we find in the strange case of nineteenth-century philosopher Friedrich Nietzsche, when his eyesight began to fail.

Nietzsche's Typewriter

Nietzsche has been described as (and accused of) many things, some of them strikingly contradictory. For example, Nazi ideologues adopted him by cherry-picking suitable Nietzschean ideas such as extreme individualism, his alleged anti-Semitism, his allegory of the *Übermensch*, and so on. And in France during the notorious Dreyfus affair, which occurred around the turn of the twentieth century, French anti-Semites vilified supporters of the Jewish Dreyfus as "Nietzscheans." He was seen by others as a skilled eviscerator of received ideas on science and its allegedly corrupting effect on knowledge; and yet more saw him as a dangerous nihilist. Nietzsche was also thought to have serious misgivings about the sustainability of Christianity in the context of Enlightenment and industrialization, and this culminated in his provocative "God is dead" proclamation in his 1882 book, *The Gay Science*. Nevertheless, his reputation survived and recovered from its incorporation into Nazidom, and today his influence, or at least readership, is widespread, without and within the academy. Nietzsche has been read and commented on by such diverse characters as Huey P. Newton of the Black Panther Movement and former US President Richard Nixon, both of whom found something (and doubtless something different) in his *Beyond Good and Evil*. In more recent times, Nietzsche societies, conferences blogs, videos, and books proliferate everywhere.

Amid all this hubbub of opinion and research into the man and his ideas, however, hardly anyone has commented on, or sought to explain, another aspect of Nietzsche: his *productivity*, and how it changed during his career because of his adoption of a new writing technology.

Consider this: Nietzsche wrote four books between 1870 and 1881, or almost one every three years, which is pretty good. After 1881, however, he managed to deliver ten manuscripts to his publisher in the seven years to 1888, whereupon he became too ill to write any longer. That was a book and a half per year, which is really good. By 1881 Nietzsche had become almost blind, an infirmity that would surely have hampered his longhand writing. How did he manage to improve his work rate? What he did was something seemingly out of character, given his views on modernity and science: he bought a typewriter. To be precise, he purchased a top-of-the-line portable Malling-Hansen writing ball, which was sent specially to him from its inventor in Copenhagen.

On the face of it, there's nothing so remarkable about this. A practiced typist can produce a page of text very much faster than can someone trying to write the same words in longhand. And as he became more used to the modern technology, this was doubtless a factor in Nietzsche's efficiency with words. But it's more than efficiency. It's more than the fact that the efficiency of the Gutenberg press was in its capacity to print page after page of the same words much faster than was humanly possible before its invention; and it's more than the fact that the efficiency of the Jacquard loom was coded into a program that could replace aspects of the ancient human role in weaving. These examples represented significant breaks with how we produce through interaction with technology. Nietzsche became more efficient. But Nietzsche was not an industry. He was a thinker. Typing transformed Nietzsche's *consciousness*—it affected how he thought about and expressed the world as he understood it. Recall Walter Ong's words on the technology of writing, how it took hold of human consciousness 3,000 years ago and changed it, with the written word representing thought itself. Written words were embedded into the consciousness of the literate to imprint a kind of thought grammar that reflected the world as they read and wrote about it.

Nietzsche's near industrial-scale productivity and efficiency came with a cost (or was it a benefit?) to how he thought and wrote before. Blindness forced him to stop writing longhand with pen and ink and instead use his fingertips to identify the fixed arrangement of letters on the Malling-Hansen writing ball. Inevitably, the grammar of the mechanical writing ball overrode the schooled grammar of longhand writing and the thought that it produced. The sudden mechanical punctuated strike of the typewriter contrasted starkly with the ruminative flow of the pen; the typewriter encouraged a binary decision, to depress the key or not; whereas the pen with its store of liquid ink, held by surface tension in the nib, or in a small reservoir in the fountain pen, was a more latent and non-machinic technology. The first is an incipiently digital form of thought expression, the second more innately analog;

one the beginning of the forming of what literacy theorist Maryanne Wolf would call the "digital brain,"[8] the other a brain formed in print culture and in the Romantic ambivalence toward science, Enlightenment, and machines.

Once he had mastered the skill of the touch-typist, Nietzsche's thoughts must have really flown onto the

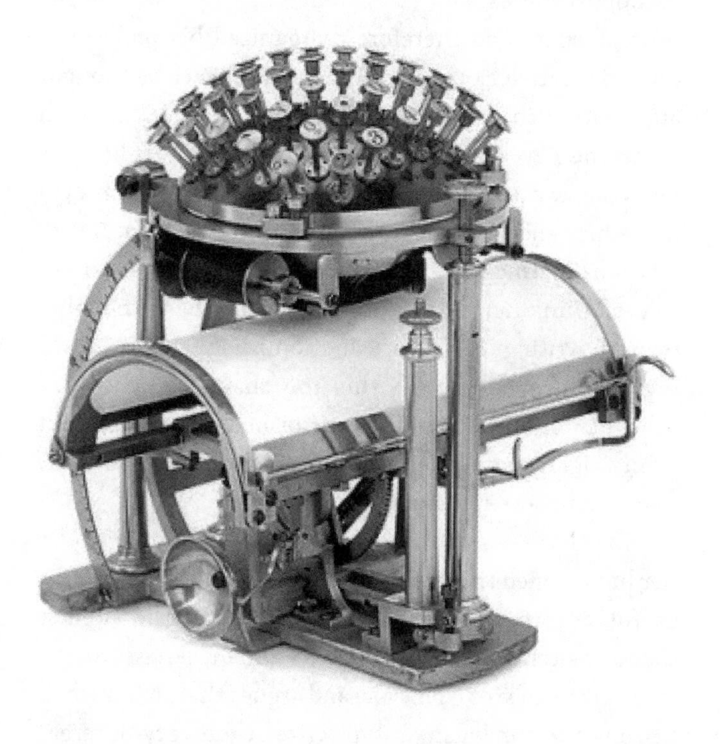

Figure 9 The Malling-Hansen writing ball, 1870

pages of typescript, enabling him to produce a new manuscript regularly in under a year. However, like the saying "for the person with a hammer, everything begins to look like a nail," Nietzsche's machine-made words, and therefore his thought, adopted a new form of technological determinism. The typewriter, with its capacities and its limits, its opportunities, and its curtailments, restructured his consciousness and therefore reorganized his philosophical and creative expression. What was able to be thought and written in, say, *The Birth of Tragedy* of 1872, when Nietzsche's eyesight still held and longhand was his written form, was something no longer possible once his eyes failed him and he got behind the Malling-Hansen. A very different writing technology helped to determine another way of thinking and the expression of these thoughts through writing. The agency and control inherent in wielding a pen, even though this too shaped and formed thoughts for millennia, was transformed when tapped out from a rigidly positioned set of mechanical keys. How was that difference expressed?

The German philosopher of technology Friedrich Kittler has claimed that the analog typewriter in general was useful for certain forms of thought: the brief, the succinct, the forms that thrive on concision and quickness. Kittler looks at the case of Nietzsche and argues that "Nietzsche's reasons for purchasing a typewriter were very different from those of his colleagues who wrote for entertainment purposes, such as Twain, Lindau, Amytor, Hart, Nansen, and so on. They all counted on increased speed and textual mass production; the half-blind, by contrast, turned from philosophy to literature, from rereading to a pure, blind, and intransitive act of writing."[9]

What would in the mid-twentieth century come to be called the "culture industry" would thrive on this modern technology and the productivity it enabled in commercialized culture, such as literature, movies, and television. But it posed an existential problem for philosophy and philosophical thought. The transformation in Nietzsche's writing due to his *écriture automatique* was so marked, indeed, that it was even noticed at the time. In 1882, the *Berliner Tageblatt* commented on the onset of Nietzsche's "complete blindness" and wrote that "with the help of a typewriter [he] has resumed his writing activities." However, the article gave notice: "It is widely known that his new work [*The Gay Science*] stands in marked contrast to his first, significant writings."[10] For Kittler, the contrast was indeed marked. He notes that the character of Nietzsche's writing and therefore his thought processes had, via the mediation of the typewriter, been significantly recast. As Kittler saw it, a celebrated Nietzschean style comprised of sustained reflection, long sentences, and complex reasoning had changed "from arguments to aphorisms, from thoughts to puns, from rhetoric to telegram style."[11] Nietzsche was an early adopter of the technology. He took it to the heady realms of Continental philosophy and—if we consider his immense influence—began to change it through a creative mind that was reshaped by the keys that had replaced the pen.

Nietzsche may not have appreciated this evaluation, but he did seem to realize in his now-sightless world that something important had occurred in his thinking processes. As Kittler again tells us, in one of the few letters Nietzsche wrote on a typewriter, and anticipating Marshall McLuhan, Nietzsche stated, "Our writing tools are also working on our thoughts." He undoubtedly was aware of his increased productivity, but as to the quality and substance of the content, as its creator, he perhaps was not best placed to judge. Kittler, for his part, was clear on what was happening with this technology at the general level of philosophic thought. Philosophy, or the thinkable and expressible elements of it via mediation, changed fundamentally with the gradually more widespread adoption of "automatic writing" and with it the culture it would produce for most of the twentieth century. Kittler noted a turning point for Nietzsche in his *Genealogy of Morals* from 1886. This book, an immensely influential volume on moral concepts, Kittler reads as symptomatic of an evolutionary change in human thought not only in Nietzsche but in Western philosophy itself. *Genealogy*, in other words, prefaced the technological evolution of machine memory in computing, which was being played out in nascent form in the action of Nietzsche tapping out his philosophy through a fixed array of letter keys. Kittler writes: "In the second essay of *Genealogy of Morals*, knowledge, speech, and virtuous action are no longer inborn attributes of Man. Like the animal that will soon go by a different name, Man derived from forgetfulness and random noise, the background of all media. Which suggests that in 1886, during the founding age of mechanized storage technologies, human evolution, too, aims toward the creation of a machine memory."[12]

Kittler's reference to machine memory is to Charles Babbage's Analytical Engine of 1835, the first computer with an integrated memory. In his design of the machine Babbage was directly influenced by the Jacquard loom, invented thirty-odd years before. Moreover, his work on computing was oriented not only to drive human error out of computation by creating a machine to do it but also to automate and computerize as much of industry as possible on the "Jacquard principle." Leibniz's written code of "universal language" would be adapted from one originally envisaged to be the common human language for sharing and developing knowledge in the arts, sciences, and philosophy to a machinic process whose logic and direction would be subsumed into the narrow needs of industry: beginning with the making of textiles, then the making of commodities of every kind, and, finally, to the ever-more pervasive making of machines by machines themselves.

The line from analog to digital in computing would run straight from Charles Babbage, who sought to industrialize the mind through machine memory, to Alan Turing, who in the 1930s sketched out in his essay "Computing Machinery and Intelligence" the formal basis of present-day computer logic. This is a line that would also run away from our analog essence and our resonances with technologies drawn from nature. However, in 1886 the typewriter that changed Nietzsche had a long time yet to cast its spell on humanity. From that time until more recently, what Maryanne Wolf called the "writing brain," the "mechanical" brain that had been formed out of the effects of the Gutenberg press, would serve to "industrialize" not only the mind but the economies, cultures, and societies that humans would erect. And this mechanical cast of mind would be the basis for the modern analog world in all its forms, especially its twentieth-century articulations with its successes and failures.

The Last Word

The printing press was the ultimate extension of a more ancient phonetic literacy. Words could now be reproduced in infinite numbers; universal literacy was at last fully possible; and words on paper became portable individual possessions that shaped worldviews. Typography, the nucleus of all modern technologies, ensured the primacy of a *visual bias* that sealed the fate of the ancient preliterate tribal and clan societies. The new medium of linear, uniform, repeatable type reproduced information in unlimited quantities and at hitherto-impossible speeds, thus assuring the eye a position of predominance over the ear, nose, and tongue. As a "drastic" extension of our capacities, writing shaped and transformed our entire environment, psychic and material, and was directly responsible for the rise of such disparate phenomena as nationalism, the Reformation, the assembly line, the Industrial Revolution, the concept of causality, Cartesian and Newtonian and Einsteinian conceptions of the universe, perspective in art, narrative chronology in literature, and Freudianism as a psychological mode of introspection or inner direction, which greatly intensified the tendencies toward individualism and specialization that had begun when we started using cuneiform to indicate ownership of goats or cereal or land or slaves. The schism between thought and action was thus institutionalized with printing. And the fragmented human, already sundered by the alphabet, was diced further into individual and increasingly smaller pieces within a growing mass society. From that point on, McLuhan's Western "man" was truly Gutenberg "man."

To be so shaped by the forces of modernity had great consequences for what it means to be embodied analog in an increasingly machine-driven and automated world,

As a “drastic” extension
of our capacities,
writing shaped and
transformed our entire
environment.

where abstract forces in production and in knowledge, the very processes that humans kept close to themselves through their hand-held tools—tools that could nonetheless build the Pyramids and the Great Wall of China, great iron and steel ships and bridges, tunnels under the land and sea, stone-built cities and continental railways—were being wrested from them by forces that were by then impossible to resist.

6

ELECTRONIC ANALOG

Electricity: “The Force of the Electrical Kiss”

In the February 1879 edition of *Popular Science Monthly*, Samuel Morse was quoted in a review of the book *The Telegraph in America*, penned by one James D. Reid, as saying: “If the presence of electricity can be made visible in any part of the circuit, I see no reason why intelligence may not be transmitted instantaneously by electricity.”[1] This was canny discernment by Morse, as we will see, who had already achieved just that feat with his telegraphic inventions earlier in the century. Writing expanded the repository of knowledge from the head to the printed page; this was the technology’s human-changing and world-changing achievement. Electricity enlarged it again, transforming the time-space coordinates of where knowledge or “intelligence” resides to the time of an “instant” and the space of, potentially, everywhere. Electricity was the valence that enabled analog technologies to step up to a new level of capacity and effect on the human and the world. To understand a little more about this process, it's helpful to know more about the historical and social effects of electricity itself, something we hardly give a thought to until the lights go out in a storm.

There was always something special and mysterious about electricity for those early discoverers and inventors who thought about and experimented with this strange natural occurrence. Strange, because unlike so much in nature, it's a force that's invisible. It's like the wind, only much less discernible in its origins, and it projects a very different power. A wind can be soft and gentle, or hard and strong, whereas electricity will manifest as a sudden jolt, a strike of energy from an imperceptible source.

Those ever thoughtful and probing Greeks were the first in Western culture to cogitate the conundrum of this unseen phenomenon. The torpedo fish, or electric ray, was noted by them to have this extraordinary power. Torpedo (like the word "torpor") comes from the Latin *torpere*, meaning "numb." In the Socratic dialogue *Meno*, Plato rather insultingly compares Socrates to the torpedo fish, as someone with a demeanor and speech that "torpifies those who come near him."[2] And many unfortunates discovered that if you touched a torpedo, the effect was immediate. The Greeks even used it to anesthetize the pain of childbirth with a discharge of 30–50 volts. But it was Plutarch (46–120 CE) who recognized, though did not fully understand, that electricity's invisibility was in fact a process of conduction. He saw that the torpedo's shock could be delivered through water and could travel a distance through a fishing pole or trident. Beyond this, however, and for all Classical Antiquity, philosophers and physicians did not really understand it.

And there the matter hung, like other Classical matters such as reason, democracy, science, and mechanics, throughout the long twilight of the Christian Middle Ages until the European Renaissance. Then, in the 1600s, various inventors and natural philosophers began to look more closely at this curious force. Discoveries, such as magnetic attraction and static electricity by William Gilbert, became widely known across Europe through the new networks of print. Gilbert also invented the first electrical measuring device, the electroscope, which was a pivoting needle that responded to an electrical charge. And it was Gilbert, moreover, who coined the term "electricus" to designate this force, a term meaning "like amber" (amber is *elektron* in Greek; amber creates static electricity when rubbed with a piece of cloth). Our familiar word "electricity" was coined in 1646 by Thomas Browne, who, incidentally, was a serial wordsmith and much venerated by Samuel Johnson for having "augmented our philosophical diction" with an astonishing 775 neologisms attributed to him in the *OED*, including, and interestingly for our purposes, "cryptography" and "computer."

With new words for new ideas that were circulating through new networks of knowledge, a renewed interest in electricity grew. Benjamin Franklin is probably the best-known member of a dazzling eighteenth- and nineteenth-century network of inventors and tinkerers who took the Enlightenment impulse "to know" down a track of energy physics that would supercharge the Industrial Revolution with a new source of power for analog machines. Franklin made some fundamental discoveries. Previously, it was believed that electricity consisted of opposing forces, but Franklin deduced that it was an "electric fire" (later known as electrons) that circulated like an invisible liquid. He described this force passing from person to person in experiments using the new Leyden jar, a rudimentary battery, where, as he explained in a letter to his friend Peter Collinson in 1747: "*A* and *B* stand on Wax; give one of them the electrised Vial in Hand; let the other take hold of the Wire; there will be a small Spark; but when their Lips approach, they will be struck and shockt." By means of these batteries, he went on to note with dispassionate objectivity, "We encrease the Force of the electrical Kiss vastly."[3]

Franklin went further to show how the fiery fluid could be generated by an "electric wheel," a precursor of the electric motor, which he invented. In this experiment, he showed how positively and negatively charged Leyden

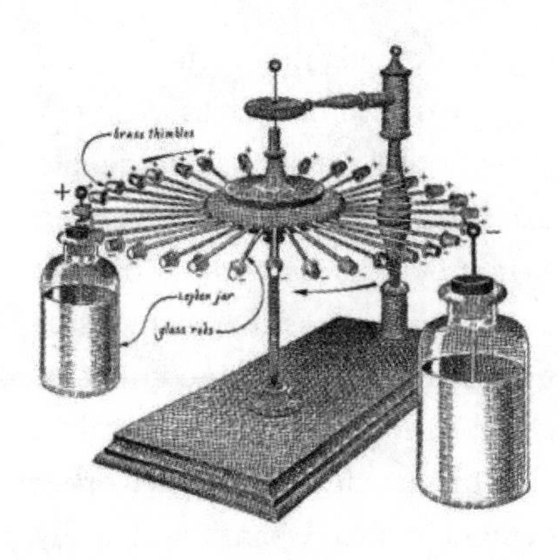

Figure 10 Franklin's electric motor, with Leyden jars, 1747

jars placed under glass spokes with brass thimbles on the end would make the wheel rotate.

The current wasn't very powerful, and it struggled to turn the wheel for very long. However, Franklin proved a concept: an electrical *engine* was possible, and it could move wheels, which meant that it could move *machines*. And ever the practical inventor, Franklin later "electrised a Machine" that, he claimed, would turn a turkey on a spit.

A dependable electric motor was the Holy Grail in what would later be known as electrical engineering. And during much of the nineteenth century, steady advances were made across the United States and Europe toward this goal. Steam power was supreme, but the industrial potential of this more mobile, intricate, and *scalable* power was clear to many. The impetus for ever more efficient and powerful machines was very strong for a class of inventors and philosophers who were also often businesspeople and/or owned shares in scientific or industrial enterprises. Franklin's maxim "time is money" was something the new business class well understood. Machines that were faster saved time and therefore money. And in the context of fierce entrepreneurial competition, a major attraction of the electrical motor, like the machine itself, was worker replacement. Factory owners realized that electrically powered machines could replace increasing numbers of workers. The more human fatigue and error that could be removed from production, they reasoned, the more would the profits flow with the current.

Electricity was in the cultural and literary air, too, as artists and philosophers wrangled with some of the same human-technology-electricity questions. For example, Mary Shelley's *Frankenstein*, written in 1818, sounded a warning about industrial science by exploring the famous debates between materialism and vitalism then going on. Materialism was rooted in religion and tradition within a world of matter and nature that were God-given and humanly manifest. The theory of vitalism, for its part, assumed an unchecked faith in scientific progress and promoted the belief that *life* was an invisible energy and electricity was nature's expression of it. In Shelley's novel, Victor Frankenstein was convinced that the Monster had been given the life force through a vitalist jolt of electricity. The tale, we know, ends badly for both Monster and Frankenstein. Shelley's moral and ethical lesson, however, was that an uncompromising faith in revolutionary science was endangering the foundations of Western Christian society.

But in the real world, business was business, and notwithstanding the moral dimension the 1820s to the 1850s witnessed a slew of breakthroughs in electrical engineering. Notable here was Michael Faraday's 1831 discovery of electrical induction, which showed how current is generated; this led to the development of the motors, transformers, and generators that underpin electrical engineering today. In 1840, Robert Davidson showed the flexibility of electrical power with working prototypes of such miscellany as a locomotive that could carry two people, an electric lathe, an electric printing press, and an electromagnet that could lift two tons. And in 1845 Paul-Gustave Foment invented the "mouse mill," a compact motor that became the basis for the telegraph.

With these and many other developments, the science around the understanding of electricity was becoming more robust. Franklin's "electrical kiss" was no longer so shocking. Nature's invisible force was able to be controlled more and deployed through industrial applications that would compete with the great parallel discovery of steam for the movement of great and small analog machines.

Both would find their place in the age of machine-making through scientific enterprise, but only electricity would have an infinite life. Possessed of the vital force that would bring the machine (instead of Shelley's human) to life, electricity would take analog technology to a new level of complexity and potential. Moreover, electrical power as the activator of all kinds of technology would, through the aegis of capitalism, find its way into almost every sphere of human experience and bring a dynamic modernity in its wake.

A pivotal analog communication technology that would help make all this possible was one that, in retrospect, was one of the simplest: the telegraph.

Telegraphy: "Everything Nowadays Is *Ultra*"

Telegraphy means "writing at a distance." In this it follows the path of the analog-human connection that began with the invention of cuneiform. And like cuneiform writing it transformed existing human affairs, setting them once more upon a new path with unanticipated consequences. The telegraph is analog in the classical physics sense in that it has relations of continuity or flow. And writing at a distance is made possible through a continuous signal that is carried from point to point over a dedicated wire. The initial objective of a working telegraph was to translate

Possessed of the vital force that would bring the machine to life, electricity would take analog technology to a new level of complexity and potential.

the electrical wave into a communication, a message to be transmitted, received, and understood. Experiments took an important step in 1816 when Francis Ronalds sent a message down a continuous 13-kilometer wire strung back and forth between two large wooden frames set up in his London back garden. By means of an ingenious alphanumeric device, the signal was coded and translated into language through a rotating alphanumeric dial at each end of the wire.

Ronalds was perhaps a little before his time. He had a vision to put an end to the "dilatory torments of pens, ink, paper and posts" and to establish an era of rapid communication through what he phrased as a network of "electrical

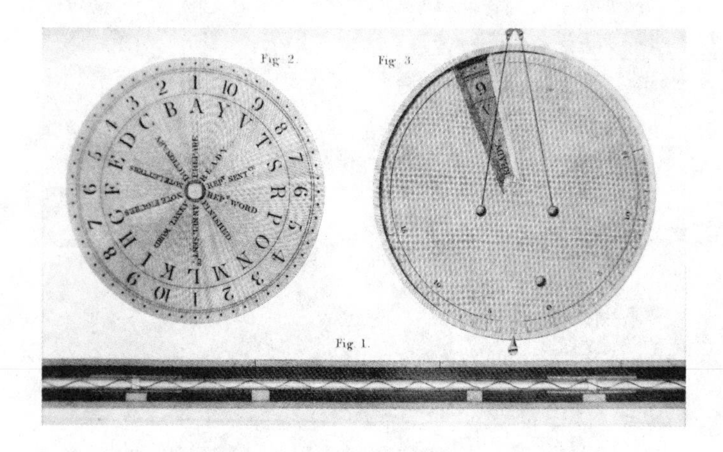

Figure 11 Ronalds's alphanumeric dial

conversazione offices, communicating with each other all over the kingdom."[4] His government sponsors didn't see it quite that way. Perhaps the administration of Lord Liverpool was too preoccupied at the time with Luddites smashing up Jacquard-type looms, or with the chaotic aftermath of Napoleon's failed territorial ambitions, or with King George III's increasingly erratic behavior. Whatever the reason, the British government took the view that "telegraphs of any kind are now wholly unnecessary"[5] and declined to fund its development.

He may have failed to realize his dream, but Ronalds succeeded in confirming a concept: writing at a distance, previously an impossible-sounding idea beyond the complicated and protracted semaphore system, was now feasible in something like real time and over vast tracts of land and sea.

Across the Atlantic Ocean the inventor Samuel Morse was faring rather better. By the 1830s the United States was a growing industrial power, and a historically unique mix of freewheeling commercial and scientific culture was emerging to challenge the hegemony of the old European powers. A growing American national spirit, coupled with a rapidly expanding continental railroad network, was conducive to the adoption of an experimental communication system that would help unite the fledgling states. Morse's invention—the telegraph system and the code that created its language—was patented in 1844. Its protocol of

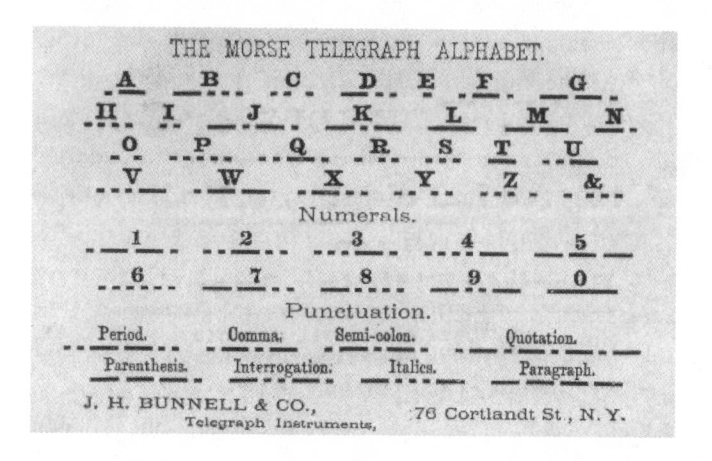

Figure 12 The Morse telegraph alphabet

dots and dashes to represent letters of the alphabet proved to be a brilliantly inspired combination to produce writing at a distance. The US Congress immediately saw the point and decided to fund his research.

However, Morse lacked the practical vision and commercial enthusiasm of someone like Ronalds. Like many of this late Enlightenment period, Morse was an artist and thinker as much as he was an inventor and scientist. Indeed, at the time Morse was better known as a painter. For all its New World dynamism, the professionalism and business sense with which the United States would one day lead the world had not yet fully permeated its educated classes. Nonetheless, the portrait painter tested his telegraph system on May 24, 1844, by sending a coded message down a wire that ran along the railway line from the Capitol in Washington to Baltimore, 70 kilometers away. Decoded, it read: "What hath God wrought?" This was an anxious philosophical question that had traveled from a split-second ago past to a soon-to-be future where the world would have changed irrevocably on its arrival—and with consequences for humanity that no one could foresee.

Such extreme shrinking of time and space by the telegraph would, when posed as a philosophical question, hew to the preexisting philosophical tendencies of the age. For many, the instant taken for a signal to travel from Washington to Baltimore was simply further proof of our species' technological genius, yet another triumph in the Age of Discovery that saw humans transform their world in so many magnificent ways. And for many ordinary people, that an invisible force could carry an invisible message down a wire in no time at all seemed nothing short of magical. That sender and receiver were invisible to each other yet could "speak" across immense distances in real time seemed incredible, but true. Newspaper adverts from the time praised this "extraordinary apparatus" as "The Wonder of the Age!!," a miracle device containing "Electric Fluid travelling at the rate of 280,000 miles per second."[6] More soberly, but hardly less impressively, on August 26, 1883, the *New York Times* reported, via a transcontinental web of wires and repeater stations that now encircled the planet, on an immense volcanic eruption on the island of Krakatoa in faraway Java in a place called the Dutch East Indies. The *Times* carried the story scarcely a few hours after it happened.

But Morse was not alone with his trepidation. Others with a similarly reflective cast of mind worried about the effects of these varied but always sensational communication technologies that were becoming a reality. The poet Goethe was one. Writing to his friend, the composer Carl Friedrich Zelter in 1825, he lamented the turn that the world was taking. The modern times, he saw, belonged to the "practical men" who were so consumed by industry and technology that "no-one knows himself any longer." He bemoaned the mania for speed that had gripped the lives of these men: "Railways, quick mails . . . and every possible kind of facility in the way of communication is what the educated world has in view [and] everything, dear Friend, nowadays is *ultra*."[7]

Another philosopher who was more upbeat, but for different reasons, was Karl Marx. In "The Manifesto of the Communist Party" he and coauthor Friedrich Engels fancied that the "electric telegraph" was contributing mightily to the first globalization of capitalism then underway. They prophesied a coming "world market," "immensely facilitated [by] means of communication" such as the telegraph, which would transcend the old barriers and the old ways and create a new and modern dynamic where, as they famously put it: "all that is solid melts into air." The telegraph, in other words, would help establish the basis for communism.[8]

The late nineteenth-century air was indeed filling with nonsolid material—in the form of electrical signals, invisible media that was working independently from the ancient and foundational connection with the human body. To write at a distance was to act at a distance, and to act remotely was to diminish the analog contiguity and resonance that had constituted the individual and social relationship with technology and nature that had endured for thousands of years. This ironic distancing through invisible connections was yet another cost imposed by industry and modernity that few then recognized—certainly not Marx, who obsessively predicted the coming of communism through new communications, and certainly not the communications magnates themselves, such as J. P. Morgan in the United States, who equally obsessively saw the telegraph and the railway as simply a way to dominate markets further, so to make more money.

What was clear was that the world became smaller for the literate millions who were part of the modern mass-media society that the telegraph helped create. And with the telegraph and later the telephone and television, the human faculties of speech, seeing, and hearing had become more technologized, meaning that individuals and society had been sensorially extended into space and time and that this gave (or so it seemed) a cognitive dominance over the world through the immense amounts of knowledge (and therefore power) that could be acquired through them.

It was the *inhuman scale* of telegraphic communications that constituted the revolutionary break with our ancient interaction with technology. Marx's phrase the "annihilation of space by time" to describe the new communications of the age meant more than New Yorkers "knowing" that Krakatoa had been obliterated a few hours previously. It was a process of shrinking that cut deep into the human-technological relationship that was built on contiguity and resonance. Recall the earlier anthropological definition of the analog relationship: that it is marked by *recognition*—the idea that analog machines are equivalencies, that they imitate actions people recognize in nature and in their bodies. We recognize the action and the extension in, say, a telescope or a bicycle. But it gets rather more difficult to recognize the material human connection when an invisible current taps out news of something happening *at that very moment*, half the world away. This technological incorporeality would become even more magical when Marconi developed the wireless telegraph in 1910.

Such lack of recognition also signaled the beginnings of the lack of interactive human control too. For French philosopher Jacques Ellul, writing in the 1960s, modern technology was a force for "the exclusion of Man [*sic*]," a

It was the *inhuman scale* of telegraphic communications that constituted the revolutionary break with our ancient interaction with technology.

force that had begun in that late Victorian age of telegraph to "obey its own laws and . . . renounce all tradition." Technology had become too clever for us; we were losing our grip, literally. Wireless communications, automated machine processes, electronic manipulation, and other advances meant that modern technologies were able to act not only mysteriously but autonomously too. The "buffer between Man [*sic*] and nature," Ellul wrote, had been removed and we could no longer "find . . . the ancient [technological] milieu to which he was adapted for hundreds of thousands of years."[9]

The sociologist Richard Sennett thought about the problem of scale and the connected problem of recognition. He wrote that an electrified machine "is ourselves enlarged"—be it telegraph, or train, or 1950s desk calculator. "It is stronger, works faster and is never tired." The problem in respect of our understanding and recognition of it is that "we make sense of its functions by referring to our own human scale."[10] The telegraph's functions and potential make little sense from the perspective of human scale. The world shrinks with the speed of the signal. Yet the telegraph made little philosophical or logical sense to its early inventors and users, anyway. So they found the illusion of recognition within pragmatism. Telegraphy worked and so was left to function at its own scale and through its own logic. Moreover, we pragmatically, if illogically, determined to enhance speed and autonomy and mystery wherever possible. Franklin's imperative "time is money" drove the process. And in the case of the telegraph, to shrink space through time for as many people as possible would be an "efficient" and therefore profitable means of communication. Such abnegation placed humanity on a new and different technological plane, one that was little understood beyond its formal scientific and economic parameters.

The material substance of our ancient relationship with technology began to attenuate much more rapidly with the invention of the telegraph. It was a pivotal analog technology that contained within it the lurking ghost of the digital computer. Its autonomous power over human communication would only increase with the coming of the telephone, television, and, eventually, the incorporation of all these into the personal computer and mobile phone. The importance of the commercial potential of the telegraph is revealed in the fact that the semi-philosophical and amateur quality of the early years of its development did not last for long. It was too important and too powerful in too many ways to be left to the passions of the independent scientist, or even to the distributive forces of the market. At the turn of the century there was still some way to go, but in the shape of the telegraph the mature Age of Gutenberg had met the fledgling form of what would one day bring forth its digital nemesis. But the new electrically powered communications media would first need to establish itself. And it did. And so the fact of the implacable logic that dictated that the twentieth century would be a media century is underscored when we recall that the first industrialized monopoly of the modern era was not a steel or oil or railroad company, as we might think, but the Western Union Company, a telegraph company.

Nevertheless, Franklin's "electric fire" needed to become more sophisticated and manipulable before electrically powered technology could take the world to another level of analog dexterity. The development of *electronics* would enable this.

Electronic Analog

The difference between *electricity* and *electronics* is not one that the public takes much notice of. Yet it is important in the history of analog technologies and is another key moment in our story. We may sum up the difference thus: electricity powers machines, whereas electronics enables them to *make decisions*.

Electrical current of the type that jolted Franklin's volunteers was the action of a circuit—a flow or wave of current that can either terminate at a point or keep traveling in a continuous loop. In physics, this energy flow corresponds to an "analogical system" that, as R. W. Gerard observed on the difference between analog and digital, is where "one of two variables are *continuous* on the other." Electronics is a different way of controlling current flow. Key is the ability to "switch" the flow of current to make it act in more complex ways and with more sophisticated effects. The breaking of the current is then a significant difference, because, to quote Gerard again, this expresses the basic features of a "digital system" where "the variable is discontinuous."[11] This ability to manipulate electricity at a shrinking scale was the basis of microelectronics—and is the platform on which digital computing is based.

Analog electronics has achieved much for humankind. The technology was literally transformative. The electrical analog technologies that already existed in the late 1940s—communications technologies especially, such as telegraph, telephone, radio, and computer—would be re-revolutionized to new levels of intricacy and capability by electronics. Indeed, supercharged by micro-electronics, the Apollo 11 journey to the moon in July 1969 was almost an entirely analog affair. At the same time, such "wizardry" (a revealing term commonly used to describe sophisticated technology) would make analog means and effects yet more abstract, detaching them further from their users in relation to control, scale, recognition, and speed. In these criteria, micro-electronics had no human analog. Electronics also enabled the automation of many more industrial and consumer technologies, something almost universally seen as progress. Indeed, some prominent thinkers prophesied that electronic automation would create a new technological hybrid, super-sophisticated robotic-computer systems that, far from alienating us, would do our bidding. For example, and taking up the "extension" metaphor to describe the human-technology relationship, the computer scientist J. C. R. Licklider wrote about it in an influential essay called "Man-Computer Symbiosis." In it he envisaged that "Mechanically extended Man [*sic*]" would "set the goals, formulate the hypotheses, determine the criteria, and perform the evaluations," whereas computers would do the all the routine work but under ultimate human control.[12] A general enthusiasm for such highly complex robotics would shape the postwar world, militarily and industrially. By the end of the 1950s, in fact, electronic systems were already producing such things as automated car factories, military command and control systems that would operate the US nuclear defense structures, and satellite communications.

It was in such an era of fast-moving and electronics-led transformation that Marshall McLuhan would think his most perceptive thoughts about the world that was then multiplying all around him. He used the extension metaphor more critically than did Licklider, however. By the time *Understanding Media* was published, mass-produced solid-state and transistor-based communication had created his "global village." The human voice and eye, he wrote, "had achieved a social and institutional extension." This signaled a new human-technology relationship, where "electric communication, also characteristic of our nervous system, makes each of us present and accessible to every other person in the world."[13] At first glance, this looks positive. However, McLuhan goes on to say that by "technologically extending our consciousness"—by expanding the "brain outside its skull"[14]—our processes of subjective experience, the most intimate means of cognizance and perception, and where the human scale takes its measure and makes sense of the world, begins to stretch toward cognitive untenability. The rupture between analog thought and action was widening with each new electronic communication innovation. Time and space and our relationship to it, through new forms of communication, became ever more difficult to grasp in our evolved analogical way. In other words, how was it possible to comprehend the fact of global communication wherein each is theoretically "present and accessible" to everyone else, at the same time? The fact is we can't, not fully anyway, so we default to a modern pragmatism and let the technology do its thing, because it can, and we can't.

McLuhan duly turns his attention to what he sees as the hidden negative: "To a large degree our co-presence everywhere at once in the electric age is a fact of passive, rather than active, experience." Passive experience—*watching* instead of interactively *doing* in the ongoing development of our eyes as the principal sense organ—was

The rupture between analog thought and action was widening with each new electronic communication innovation.

to cede control but in a way that was masked by the illusion created by the magic of instant connectivity. This was the most important effect of electronic technology. McLuhan's "message" was that automatic decision-making at the micro-level of whatever technology electronics was applied to, coupled with automation at the macro-level of the technology's action upon the world, served to doubly alienate us from its action. The electronics revolution was therefore an effecting of what Michel Foucault would call "docile bodies," or "subordinated cogs"—within a society-wide state of "automatic docility."[15] But as is a common feature of technology history, this was a process of subjugation that went largely unnoticed due to our collective investment in the ideology of technological progress.

The widening range and power of electronic analog technology was changing not only industry after industry in the 1960s, but also, and unavoidably, the cultures and societies that had long since become industrialized themselves. The "global village" was where a good deal of humanity now lived at this time: in North America, in Western Europe, and in its own way, in the communist bloc countries where science and technological know-how was also advanced, even if consumer goods production wasn't. Satellites made the global village possible in large part. And satellites themselves, hitherto avatars of science fiction, had captured the popular imagination of the world, especially after the Soviets launched the Sputnik satellite in 1957. Sputnik triggered the "space-race" between the United States and the Soviets, but it also prompted a commercial battle within the West to continue the communications revolution by bringing increasing numbers of connected consumers into the global village. For example, a US, French, and British consortium launched the first Telstar satellite in 1962 to transmit telephone, television, and telegraph communications.

Figure 13 Model of the 1962 Telstar satellite

Telstar was a breath-taking technological feat that was set to "reshape Man's [*sic*] future," as a newsreel put it at the time. And a watching global public was duly captivated and inspired by the accompanying propaganda, especially from US government and news media that had been stunned and shocked by the Soviet success of Sputnik beating them into space—so much so that Telstar and its space-age ethos achieved a massive cross-over into popular culture. For example, the British instrumental pop group the Tornados released a single called "Telstar" in the same year as the satellite launch, and with it became the first British band to have a number-one success in the United States. Indeed, the song, produced by experimental sound engineer Joe Meek, was a huge hit across the Western world. A major element of its sales triumph was its distinctive high-tech sound, a sound previously never heard by most people up to that time. The songs' futuristic melody was made by a "Clavioline," an electronic keyboard, invented in the late 1940s, that used switches to alter and control the tone of a continuous vibrato that could be distorted by the keyboard player. With "Telstar" electronic music was launched to a global audience.

Another famous electronic keyboard was the "Mellotron." Introduced in 1963, this organ was based on sound tapes that ran in continuous loops inside the body of the machine, and it inspired musicians to explore the popular space-age themes pioneered by the Tornados and their sound engineer. Recordings such as David Bowie's "Space Oddity" (1969), Pink Floyd's "Celestial Voices" (1968), and various parts of *Dark Side of the Moon* (1973) all used its sounds. The Mellotron's owl-like hooting notes featured also in the famous intro to the Beatles' "Strawberry Fields" (1967). Staying with popular music, electronic transistors had transformed the 1940s pick-up electric guitar into what was now strictly speaking an *electronic guitar*, though no one ever called it so. Guitar innovations such as the Fuzz Face and wah-wah pedals meant that sound waves could be controlled and manipulated to make, again, sounds never heard by the human ear. Indeed, these pedals produced something of a signature sound of the late 1960s psychedelic era when Jimi Hendrix gave a master class in both Fuzz and wah when headlining Woodstock in 1969.

By the time Woodstock came around, television news anchor Walter Cronkite had been for years telling Americans, "That's the way it is" in respect of the reported events of the day. And for years, since the early 1950s in fact, television had been asserting itself with the masses as the most desired electronic technology. The combination of vision and sound (remediated film and radio) proved a powerful stimulant to senses hungry for what this new technology offered. Despite its inherent analog characteristics of variable picture quality to due to interference, or "snow," produced by electromagnetic noise attracted by the antennae, television was instantly and wildly popular, with ownership in the United States already at 85 percent by 1958. Satellite delivery of television and, later, color television, which appeared in the mid-1960s, saturated the senses and enhanced the passive experience of communication yet further. Through television, especially in news and current affairs, viewers watched a narrowly focused and edited view of an abstracted wider world that could often be drastically at odds with the actual world that surrounded them in their homes and towns and cities. Such passive or semi-passive watching affected cognition and psyche by consuming content that was basically an illusion, what McLuhan called a "heightened human awareness" of the world based on nothing more than a flickering or blurry analog signal from an unseen source that conveyed, once more, the effect of something magical to many of its watchers.

Television produced an analog *flow* of information and knowledge, just like, and in their own ways, the telegraph, film, and radio before it. Moreover, and like these previous forms, the stream of sound and vision was disembodied from direct apprehension and was both mediated and immediate. This placed the viewer at a disadvantage as compared with the cognitive interaction with previous major analog information technologies such as the book. Electronics makes the flow of television, from production to consumption, much more manipulable and controllable from the source. This is important. Such technical efficiency, a main criterion of which is commercial efficiency, or the capacity attract and retain the viewer with a steady stream of watchable and compelling stimulation, means that there is little or no time to properly reflect on what is seen as information and knowledge, be it a 20-second advert for toothpaste, an hour-long version of Shakespeare's *Twelfth Night*, or a three-minute news story from Walter Cronkite about "pacification" measures in Dinh Tuong province in South Vietnam.

By contrast, words on paper, such as in a newspaper, magazine, or book, were fixed in time and space. The reader could take their time and read closely and reflect on the information. This was the practice of a print culture that had existed since the time of Gutenberg and had shaped our knowledge of, and interaction with, the wider world. This was being dramatically destabilized by the electronics of the new medium—and we barely noticed.

The cognitive effects of the electronic flow of television sounds and images was studied by Raymond Williams, a founder of the cultural studies discipline. In his 1974 book *Television: Technology and Cultural Form*, he noted in his analysis of TV news programs the lack of connection between clearly related items, and the editorial effort to fuse all program items—advertisements and features, related or not—into a sequence or "flow of consumable reports and products"[16] designed to function as a commercial

Figure 14 Watching the news in 1968

package. Granular electronic control over the film or video editing, transmission, and reception processes allowed for the creation of a particular kind of broadcast communication, one that was profoundly different from print culture communication. As Williams puts it: "In all communications systems before broadcasting the essential items were discrete. A book or a pamphlet was taken and read as a specific item. A meeting occurred at a particular date. A play was performed in a particular theatre at a set hour. The difference in broadcasting is not only that these events, or events resembling them, are available inside the home, by the operation of a switch. It is that the real programme that is offered is a sequence or set of alternative sequences of these and other similar events, which are then available in a single dimension and in a single operation."[17]

The electronic power to refine, control, and automate what analog technologies could do, especially communications technologies—and in this we need to include transportation technologies such as jet aircraft, space probes, satellites, and rockets taking men to the surface of the Moon—took the human-analog relationship to its highest level of sophistication and to its furthest extension of *distance* from the original hand-eye relationship of resonance. The technologies that so shaped our world at this time were losing their analog essence as it pertained to the human creation and use of them. The shrinking of time and space through transportation and media technologies now represented the world differently to us, and so we were forced to come to terms with the new reality it symbolized and to try to understand it accordingly. And in a precursor of our present age of misinformation, disinformation, and "alternative facts," late electronic analog media technologies were sophisticated and complex enough to produce a real-time global village that encompassed billions of people in shared experiences of a representationalized world, a world of primarily visual culture and a world contrived primarily as spectacle.

Guy Debord, a 1960s revolutionary and mordant critic of the televisual consumer society, saw alienation as the fundamental effect of new technologies in the service of passive consumption. What would on its own be an intolerable level of alienation, he argued, was leavened by the very act of consuming itself: the buying of "things" that would fill the psychological space opened up by alienation. Alienation was tranquilized by the spectacle, constituting the circus that would come with the bread. And it was the spectacle, distant yet near, that began to permeate life and make itself the touchstone of what constituted the real in the service of consumption. In *The Society of the Spectacle*, Debord writes: "In societies where modern conditions of production prevail, all life presents as an immense accumulation of spectacles. Everything that was directly lived has moved away into a representation."[18] He blasted the ethereal and insubstantial form of this reality: "For the present age, which prefers the sign to the thing signified, the copy to the original, representation to reality, the appearance to the essence . . . illusion only is sacred, truth profane."[19]

The popularity of the technology suggests that ordinary television watchers didn't see it that way. Its spectacle-producing capacity and its questionable depiction of reality quickly came of age in the 1960s. For example, a majority of those polled saw the first televised Nixon-Kennedy presidential debate in 1960 as an objective deliberation between two aspirants who appeared as "real" people before them. That Nixon was in fact unwell and had refused to wear makeup to improve his image were facts unknown to the viewing millions. His sweaty and pallid physical representation were read as signifiers of ill-health and weakness—and spelled disaster for his candidacy. Another poll of those who heard the debate on radio found Nixon had won with 49 percent compared with 21 percent for Kennedy, while 30 percent who watched on television said Kennedy had won compared with 29 percent for Nixon.

The first global spectacle for the global village was the Olympic Games. Millions "witnessed" the decade's trilogy of Olympic spectacles televised from Rome, Tokyo, and Mexico City, respectively. Overtly spectacular opening and closing ceremonies were choreographed and staged to promote "a peaceful [global] society concerned with the preservation of human dignity" under the five interlaced rings invented to symbolize the "Union of the Five Continents." But away from the television and the passive consumption the objective world of the 1960s was one of Cold War, hot wars, revolutions, assassinations, civil rights struggles, and ideological and political convulsions, either or all of which could join in with the flow of the televised Games to produce "single dimension . . . in a single operation" for the networked evening news. The 1972 games in Munich were set to follow the same Olympic script, but another reality unexpectedly broke through with the taking of Israeli hostages by Palestinian terrorists. However, the power of television to commodify, integrate, and spectacularize soon asserted itself, and the murder of the hostages and the killing of the Palestinians by the West German authorities became a global televisual news drama—one watched on television by the terrorists themselves before their deaths—and was manufactured in real time into the global spectacle of the year.

The "way it is" was the way the technology of television presented reality as a flowing and seamless commodity. The one-way technology of television created a world that we watched, and by watching it, we were necessarily not part of it. This cosmic distancing had its own cosmological representation in a single, but immensely famous, photograph taken in 1972. A crew member of the Apollo 17 spacecraft shot the first whole-Earth photograph to be taken from space by a human being. The image became an icon, a spectacle of an Earth that humanity watched taken from the perspective of a place it has never been. In that single image, representation, commodity, illusion, and distancing intersect in a 1/250-second shutter-speed instant into a single analog focus, a single image of a planet that we had been the physical children of for thousands of years. However, momentous advances in technology, culminating in the sophisticated electronic expressions of analog technologies, had finally and decisively distanced us from our Earth Mother—symbolically as well as physically.

A Coda for the Analog Age

The 1960s counterculture (counter, among other things, to what was perceived as a mainstream culture addled by violent and commercial television) and its relationship to technology, and especially to computing, becomes important at this point if we want to understand the historical transition from analog to digital. Contrary to the popular notion of the 1960s being the psychedelic era of "turn on, tune in, drop out," many saw the fracturing of the immediate postwar social consensus as proof that popular and democratic technological change, engagement with the actual material world, was where the opportunities for individual freedom lay.

American architect and inventor Buckminster Fuller in the 1950s talked about "spaceship earth" and about a world of problems that could be solved by a "system" approach that would be shaped and guided by a global "master computer." Fuller's influential but basically sci-fi visions were brought down to earth by more practical counter culturalists, especially those from around the San Francisco Bay Area in the United States. These included Stuart Brand, who founded and edited the *Whole Earth Catalog* in 1968. Brand was part of a counterculture that saw sustainability as being undermined by a rampant and wasteful consumerism propagated by an all-powerful television culture. The journal's subtitle was *Access to Tools*, and on its covers were various photos of Earth taken from space. It listed for sale such things as books and work gloves, and featured articles such as how to make a tipi or a solar panel, and the ecological benefits of Japanese architecture. Mostly, the emphasis was on items that were:

> Useful as a tool,
>
> Relevant to independent education,
>
> High quality or low cost,
>
> Not already common knowledge,
>
> Easily available by mail.[20]

While there was much counterculture rhetoric in the *Catalog* about needing to reconnect with the Earth and with ancient and new knowledge, there was also an early focus on the benefits of computers (as part of a Fuller's "systems" approach) as freedom-enabling (and still-analog) tools. In the first issue, cybernetics was promoted through a listing for Norbert Wiener's 1948 book *Cybernetics: Control and Communication in the Animal and the Machine* (for $1.95). A blurb for the book, presumably written by Brand, praises "McLuhan's assertion that computers constitute an extension of the human nervous system," but this, he cautioned, "is just one aspect of these new understandings about communication." It concludes: "Society, from organism to community to civilization to universe, is the domain of cybernetics."[21] The entry quotes from the book itself, where Wiener emphasizes the importance of cybernetics for self-organizing systems, especially to promote stability in a disorganized system.

Self-organization theory is key to understanding the changes taking place at this time. It was tied up with automation as the core of cybernetics. For the early enthusiasts it followed that computer systems (analog or digital) would help bring a presently disorganized world together once again. And it followed that the more automatic the system, the better. In the social world, networks of individuals within networks of computers would enable a return to the state of homeostasis that Fuller, Wiener, and Brand thought to be the natural order of things. In this way "Machines of Loving Grace" would "watch over us," as the San Franciscan poet Richard Brautigan put it, writing during the 1967 Summer of Love.

The mock funeral for the hippie movement that was staged on October 6, 1967, in San Francisco's Haight-Ashbury district symbolized more than the end of the Summer of Love and the dawning of a new post-psychedelic day. In the same city the communes were already packing up and making way for computer clubs. The idea of popular computing had caught on, especially in Southern California, and individuals there and beyond were doing DIY computing, using the tools that

Figure 15 Production process for the *Whole Earth Catalog*, 1971

were becoming commercially available. For example, the famous Homebrew computer club, which began in 1975 in Menlo Park, is credited as influencing early Silicon Valley leaders such as Steve Jobs and Steve Wozniak. The Club's ethos stemmed from the People's Computer Company, also from Menlo Park, whose 1972 newsletter announced: "Computers are mostly used against people instead of for people; used to control people instead of to free them; Time to change all that—we need a . . . Peoples Computer Company."[22]

The meaning of the "PC" acronym has been lost to the congested history of post-1970s computing—especially for younger generations for whom it means a computer that was not Apple. But, of course, it meant *personal computer*. And we forget also that this was an astonishing thing, something once more seemingly magical: a computer that you owned personally at home or used in your office to do all kinds of new things. The "personal" signaled (or promised) freedom for the individual. For Silicon Valley marketers, freedom was freedom from a particular constraint—from the bureaucracy and rules in business and in government. For example, the celebrated Apple Macintosh advert "1984" promoted the personal computer overtly as a technology of freedom from an overbearing and Orwellian state. And with ironic reference to Orwell's "telescreens," television was the vector for this ad, broadcast first during US television prime time, and subsequently to become the "most famous Superbowl ad of them all."

The irony—if that's what it was—didn't matter. What Silicon Valley was selling was its own distinctive idea of libertarianism, distinctive in that it claimed that freedom was possible not through laws, or the writings of a particular philosophy, but through the application of computer systems to your life. What they were selling was *computation as technology of freedom*. That's how radical it was. Such libertarianism was met with skepticism by some. But mere skepticism was never going to stop an idea whose time had so evidently come. And come it did. Intertwined with the neoliberal revolution in business, and capitalism more broadly, the 1980s and 1990s saw the personalization of computing not only in business—as it always was a business tool first—but in culture and society too. The rise of what would become the Internet, or something like it, was therefore inevitable.

Among all this objective change, what was hardly noticed was that as our lives became more computer-driven and networked, and as the logic of digital began to touch and change technology after technology and process after process, our subjective relationship with technology began to change. The *nature* of digital and its logic was lost on the average user who would experience mainly amazement at what a computer with its processing power could do, and what more it could do when connected to other computers. And this transition to a digital world—for that is what it is—was hardly reckoned by most people in its philosophical and anthropological aspects. Gutenberg's printing press took a couple of hundred years before its cognitive-cultural effects began to change Europe. The speed of the digital-analog transition was hardly more than a couple of decades. Such accelerated change was a major factor contributing to our collective ignorance about what the revolution meant—and what the subsequent obsolescence of much of our analog essence and relationships signified.

7

ANALOG TO DIGITAL

Who will be man's successor? To which the answer is: We are ourselves creating our own successors. Man will become to the machine what the horse and the dog are to man; the conclusion being that machines are, or are becoming, animate.

—Samuel Butler, "The Notebooks of Samuel Butler," 1912[1]

Every piece of information in the world has been copied, backed up, except the human mind. The last analog device in a digital world.

—Dr. Robert Ford (Anthony Hopkins), *Westworld*, HBO Series 2, Ep. 7 (2018)

A future where the boundary between carbon-evolved and silicon-built life becomes porous is approaching us at warp speed.

—Christof Koch, *The Feeling of Life Itself*[2]

The Analog Perspective: A Short Recapitulation

I've sought to impart essential knowledge underpinning a functional understanding of analog. But the book does more. It does something different and something necessary within this field of knowledge and at this time in history. Although technology is an important aspect of this book, the scope went beyond technology. Through a combination of philosophy of technology and anthropology, the essential knowledge here incorporates the living and evolving human within the dynamic and evolving connection with technology.

My approach signaled a desire to view analog holistically, to widen the focus so as to contextualize it in such a way that an understanding of it also becomes an understanding of what it is to be human in the world, in the past, present, and future. This is especially necessary in our new age of domination by digital.

This methodology was supported by a simple organizing thesis drawn from two thinkers who in merger give insight into the human-technology interaction. First was Arnold Gehlen. His "philosophical anthropology" of the 1950s built on the earlier "German School" of technology pioneered in the 1870s by Ernst Kapp. Second, and rather more celebrated, was Marshall McLuhan, whose media philosophy changed our understanding of the importance of media as shaper of the social world. Both sought to understand technology through a philosophy that can be summed up in my paraphrase: *As humans we invent and shape our tools, and these reinvent and reshape us in an ongoing interaction*.

The technologies McLuhan refers to, although he was never explicit about it, were analog. It probably never occurred to him to be explicit because when he wrote, and for many centuries prior, analog had the field to itself. From the slingshot to the missile and from pictographs to satellite television, the basic property was analog. To the extent that we even considered their nature, most technology was analog in that it was recognizable to us in its elemental workings, and it resembled the workings of our bodies and/or the natural world around us. The extension principle illustrates this. And so, we can understand that with, say, the invention of the wheel (a recognizable extension of our feet) we and our world changed and evolved extraordinarily, albeit over a long historical durée. Similarly, with the invention of writing (as a recognizable extension of the spoken word) we can see how the process of mutual evolution led from simple cuneiform accounting systems to a global mass media system based on satellite communications, this time in a much faster process of historical change.

McLuhan declared that the medium is the message. But even the celebrated media philosopher had, in retrospect, a too-narrowed perspective. This is understandable.

McLuhan did not seek to investigate the *nature* of the relationship between humans and technology other than to argue that it was interactive and that it evolved. He argued "only" that *we change* in the evolving relationship, but in this he *assumed* that the technology itself would be analog and that the interaction is based on the ancient equivalency. Thinking holistically and using insights from anthropology and philosophy, it follows from McLuhan, Gehlen, and others that *we ourselves are analog* of the earliest tools that we developed—and in their form, function, and logic, these are analog of us.

It's only now, in the context of digital, that we can more fully understand analog. And we can also understand digital better in the context of the analog it has vanquished. It was a transition that occurred rapidly. Many who lived through it were oblivious to its import, not least because of the accompanying ideology of progress that encouraged the belief that computerization was a solution looking for a problem. There were dissenters, however. In 1985, Bernard Stiegler, a German philosopher of technology, wrote, "A line was crossed when [our] brains were made to operate with digital information, without analogy with their origin."[3] The problem for modern society was the privatization of knowledge (as information) that was being sequestered within commercial databases to be stored or sold as commodities. Another problem was the ceding of control of such knowledge, not only to the proprietary servers of corporations but to the machines themselves, functioning fundamentally as autonomous actors, as slaves not to humans but to the self-learning algorithms that drive them.

Crossing the Line

Let's begin by correcting, slightly, Stiegler's statement. Maintaining that "a line was crossed" suggests, as does much of the ideas of modernity that he so excoriated, that society is (was) slouching toward progress. And writing at a time of general enthusiasm for the coming personal computer revolution and seeing digital as a future we were moving toward, it seemed an understandable turn of phrase. Today, however, we can see that the direction of travel was the other way around, with digital and the future it contains moving toward us. "We" weren't crossing a line, so much as digital technologies—computerization—was colonizing our personal and social spaces, armed with promises of a bright future if we adapted to the needs of what was (and still is) a business revolution.

That the revolution was coming toward us, the ordinary members of society, was recognized by some, one as early as 1967. Then, Robert MacBride argued in his book *The Automated State* that computers constituted a "new force in society," one that has the capacity to "reach out" to analog machines and people to shape them to its logic.[4] This "force" was held in check by governments and unions wary of the job-destroying capacity of automation. By the mid-1980s, however, with small-government and free-market ideology beginning to dominate in the Anglosphere, in particular, such fears were forgotten as the needs of competition and the imperatives of profit were ascendant. Carried by powerful winds of economic change, the "line was crossed" as computerization and automation duly occupied large swathes of economy, culture, and society—as well as the private realm of the individual. Our bodies and brains were indeed "made to operate with digital information," but in ways and with effects that were barely considered beyond a soon-to-be-hegemonic Silicon Valley–inspired boosterism.

For those who did consider the transition to digital, most drew from a Whiggish liberal culture that viewed modern technological development and modern capitalism as fundamentally positive. Problems or unexpected consequences such as global warming from rising carbon dioxide emissions, or the carnage of road death by automobile accidents, would be eventually worked out through the application of reason and improved technology. However, Sigmund Freud, in his *Civilization and Its Discontents* of 1930, had already analyzed our complacency toward technological development and the generally unchallenged belief in progress that went with it. His diagnosis was that

"Man [*sic*] has . . . become a prosthetic god," meaning that by extending the body and mind through new technologies, we have come to see ourselves as masters of the world. Freud conceded that we sometimes can reflect that such power may cause "trouble at times," but, he continues, such technologically induced trouble is seen usually by us to have a technological remedy, so that "when he puts on all his auxiliary organs," man believes himself, arrogantly and deludedly, to be "truly magnificent."[5]

Freud's description of our deep psychological investment in modernity and progress fits with the positive narratives that illustrate the major analog extensions that we have discussed in this book. Writing, the printing press, telegraphy, electricity and electronics, satellite communication, and even the Moon landing were all "truly magnificent" achievements made possible by these "auxiliary organs." Nevertheless, the negatives that underpin them—a broiling planet and 1.35 million annual road deaths, to name but two—are not so easily smoothed over. What economists term "externalities," or the hidden social, environmental, and cultural costs of progress, are baked into late modern capitalism. Automation is another such hidden externality. In respect of our interaction with technology, an ancient and deeply rooted human beguilement with automation obscures a fundamental problem that is generated by it. As Gehlen put it: "The fascination with automatisms is a prerational, transpractical impulse, which previously, for millennia, found expression in magic—the technique of things and processes beyond our senses."[6]

This "impulse" has never left our species. Automation, especially after the Industrial Revolution, hovered as a potential over almost all new inventions in machines in the production process.

And as I've pointed out several times, we've never freed ourselves fully from the impression that devilishly complex technologies are a kind of magic. A diminishing direct physical involvement with technological processes stems from the fact that with every major development in automation, from the Jacquard loom to cloud computing, we become correspondingly unable to assert control over the direction, effects, and externalities of the technologies deployed into the world. The present-day process of a generalized automation made possible by digital technologies indicates that no longer do we shape tools and are shaped by them in turn, but that it is we who are being shaped in an increasingly one-sided process. In other words, the transition to digital has seen an extraordinary deterioration of human control and participation in the human-technology relation.

It is only now that are we are coming to realize, albeit vaguely, that this colonization is also an open-ended "decrease" of "practical agency" as Brett Frischmann and Evan Selinger phrase it in their book, *Re-engineering Humanity*.[7] They argue that we "cede control over our desires and decisions" to automated algorithmic methods that we do not understand, and which find no equivalences in nature such as we find with analog technology. This affects us in two main ways. One is *physically* in respect of our relationship to machines and the material world, and the other is *cognitively* in terms of how consciousness and the synaptic architecture of our brains are being reshaped by a world of digital representation displayed on screens. Let's look at an example of each in their turn.

Analog to Digital: A Physical Effect

Ostensibly, the objective of automation is clear—it is to relieve body and mind of the burden of labor. And this seems to be a reasonable and honorable principle. As humans we are innately constrained by what our bodies and machines can and cannot do, and if automated actions can perform labor, then not only will they release the individual from exhausting and life-shortening work but they could change society in positive ways too. As we touched on before, Aristotle understood the principle of automation long before the practice was possible. In *Politics*, he observed that when machines run "of their own motion," then "chief workmen would not want servants, nor masters slaves." His thinking suggests that the more complex the machine, the more extensive and intensive the positive unburdening. For centuries that has been the general assumption. And in any case, what could go wrong with the saving of human energy and our release from drudgery and with the freedom to pursue one's own ambitions and to fulfill any talents inhibited by the toil of manual labor?

A good deal can go wrong as far as colonization by digital is concerned. As automation and computerization ripple further outward, and as these machines appropriate from us task after task in job after job, and as artists and craftspeople disappear as a distinct class or group

Figure 16 Humans and robots working together in a Japanese production plant

to become niche workers for expensive specialist markets such as printmaking, watchmaking, glassblowing, beermaking, winemaking, furniture-making, and various kinds of handcraft, then something deeper and more universal is lost: the idea of *Homo faber*. This corresponds to the earlier discussion about the growing realization by writers such as Robert Pirsig, Matthew Crawford, and Richard Sennett that we are losing touch with the analog practice of working with our hands, of making and fixing and tinkering with machines and technologies that we can recognize and understand because we evolved with them.

We can see such loss sociologically. For example, a weekend routine for millions of (usually) men was to spend hours under the hood, tuning, replacing parts, and generally maintaining a car whose temperaments and proclivities they felt they knew through the experience of interaction with them. Today, even modest cars are now highly complex machines, intricately constructed and controlled by electronics and computers to function as efficiently and automatically as possible. Most people have no idea how to work on these machines beyond topping up the oil and water. In the 1970s it was said (possibly apocryphally) that if you bought a British Leyland car that had been assembled by human hands on a Monday, it would be a lemon. The workers on the assembly line, so the story goes, would likely be suffering from Mondayitis, or be hungover, and their human foibles, the human expression of their being shaped by our ineradicable crooked timber, will have been mentally and physically transferred to the ill-fated product. Cars today are produced 24/7 to rigorous and complex specifications by robots that execute programmed tasks constantly and unerringly and with an exactitude and energy that no human could ever match. Fully driverless cars are set to remove the final element of physical and cognitive agency from this key (and once analog) technology of modernity.

Automation alleviates the burden of routine, tedious, and physically taxing work—and not only in automaking, but across the entire economy, from shelf-stacking to crop-harvesting to bread-making to garbage collection. Businesses, in the main, will give over anything they can to roboticization, and be given a tax break in many jurisdictions for doing so. Accordingly, commodity production has given way to the production of services as the major form of employment, services such as accountancy and teaching and even surgery, which themselves are subject to the same predatory forces that seek to erase the human element in production wherever possible.

A significant effect of rampant automation is one we're acutely aware of. Discharged of the need to perform manual labor, the workforce is becoming *sedentary*. The reduction of physical activity in the production of goods has been going on for a very long time, but the process has accelerated greatly in the past couple of decades due to computerization. We're acutely aware, too, of a consequence of decreasing physical activity: obesity. The mechanization that produces cheap labor is the same process that keeps many of us physically underactive. Hundreds of millions of us consume more calories than we need and burn less than we should to stay at a healthy optimal weight. The World Health Organization tells us that obesity has tripled since 1975. It estimates that in 2016, more than 1.9 billion adults were overweight. Of these, over 650 million were obese. In 2020 the WHO calculated that 4 million people per year die from the effects of obesity. This compares with the 3.5 million 2021 death toll from the global pandemic of the early 2020s that so convulsed the world. And obesity was implicated heavily as one of the underlying conditions that made people vulnerable to coronavirus's worst effects.

A burgeoning obesity death rate correlates with the rise of the digital-service economy, suggesting that the conjunction of automation and human (non-)physical work is literally lethal. It's a baleful integration depicted in the animated film *WALL-E* (2008). In this Disney dystopia the Earth has been trashed by humans and our species has long since been evacuated by a megacorporation to cruise the universe indefinitely, living in an interstellar starship where all their requirements are catered for through automation.

Figure 17 Still from *WALL-E* (2008)

Their lives are so automated, indeed, that all they need do is watch screens and consume in a strange moneyless and competition-free consumer economy. The protagonist, WALL-E, a "load lifter" robot, ironically, is the only one with the capacity for reflection and recognizable human feelings. Humans have traveled far beyond simple alienation in the film and exist in a free-floating haze of consumption and distraction. Our connection to technology as a coevolutional process has been lost, and in fact human evolution has stopped in this hyper-automated non-world. Life has no point beyond existence and existence is never contemplated, because what would be the point in a life of pre-programmed experience?

WALL-E is science fiction but its relevance as portent is clear. Super-automation causes us to lose the capacity to act as *Homo faber* and in this we lose the resonance with, and recognition of, tools that give us agency and potential in the material world. We're nowhere near the film's dystopic vision, but every day we move in that direction as we cede sovereignty and control to processes of automation and computerization that few of us asked for, but became quickly dependent on, like the mobile phone and its burgeoning app economy, Wi-Fi connectivity, online shopping, social media, instant messaging, and the myriad fruits from an invisible tree of automated production and services that magically drops life's necessities onto our laps. The film poses ultimately philosophical and political questions concerning knowledge and choices where technology is concerned. I'll come to the political issues shortly, but to conclude here, I want to frame the philosophical issue of bodies and technology through what the philosopher Bryan Magee has this to say about the issue, in a world, he tells us, that we barely comprehend: "We build purposes into our machines whose functioning and output are such as to be *intelligible to our senses* and our minds. . . . All the forms and categories in terms of which we perceive or conceive anything at all, with the aid of no matter what technologies or theories, are dependent for their ultimate intelligibility on the nature of our bodily apparatus, which is contingent."[8]

Our bodily apparatus is, in the technology relationship, certainly contingent, but that space of contingency has been colonized by digital technology, and the transition to a digital world is driven powerfully by its in-built mission to expurgate humans wherever possible. By collectively consenting to that, we have unwittingly unmoored ourselves from the anchors of analog technology and the natural world. We live a life of always-accumulating information consumption, represented on pervasive digital screens, where the critical analog qualities of resonance and recognition diminish in a zero-sum process. The more we embrace (or are embraced by) digital technology, the more we lose our analog roots and heritage. In the transition to a new category of technology, the logic of the path is therefore toward something like the dystopian world depicted in *WALL-E*. But contingency is also the existence of possibility. Our society could go another way, but we can leave our pre-programmed path only if we recognize and understand what we are losing on the forced march to digital.

Analog to Digital: A Cognitive Effect

It feels so natural that we need to remind ourselves that to write and read is to interact with a technology. At a deeper level we need to remind ourselves also that the

The more we embrace (or are embraced by) digital technology, the more we lose our analog roots and heritage.

technology is analog, in that it corresponds, symbolically, to speaking, to hearing the voice and, with it, the mind's thoughts. Writing and reading fills our consciousness with the content of what the media generates by way of aspirations, dreams, plans, phobias, emotions, fears, and other thoughts. Anything that "comes to mind" comes to it as cognition—as knowledge and understanding—and so much of this is acquired via this astonishing technology that literate persons take for granted. The technology orders mind and consciousness too. It trains it toward linearity and a form of analytical thinking that correspond to the ways in which the patterns of words on paper, written or read, are organized and structured as texts.

This technological structuring and linearity give shape and expression to the ordering of the mind. It's called narrative thinking and is something that we learn not only through literacy, but also by means of the literate cultures into which we are born. This is constantly reinforced through the ways that we communicate with each other in speech as well as through writing and reading. Narrative structures are based around stories. And these, generally, follow a linearity of beginning, middle, and end, and in that sense are analogical. What are called "meta-narratives," or overarching and legitimizing narratives, we see, for example, in the biblical stories of Genesis through to Revelation, in the foundational structures of literature from the early modern period to today, and in the legitimation stories that have shaped and guided the collective consciousness of the Western Judeo-Christian mind for two thousand years—such as the idea of Enlightenment, of the majestically unfolding ideas of the "truth" of science, and the intrinsic justice of democracy and the path toward human progress that these supposedly take us down.

Seen in this way, writing and reading is a powerful interactive technology. It gave us belief systems to hold to, the concepts of knowledge and truth, and the idea that science and democracy begat progress and justice. All this and more from a technology consisting of abstract symbols on paper that were eventually mass-produced through the Gutenberg printing process and all the subsequent improvements and refinements to it. Humble marks on paper, but they placed us on a specific path, by way of a specific technology, to produce a specific (though internally diverse) human-technological interaction that produced what would become a Western-dominated, and typographically produced, *print culture* that was expressed in myriad local ways, but expressed eventually through an overarching *modern consciousness* that gave us almost everything that expresses what we are as conscious beings, from metaphysical beliefs to materialist certainties to philosophical doubts.

So, what about the process of writing and reading itself? How has the transition to digital affected this?

Figure 18 The Gutenberg Bible (Taschen)

Words on paper are fixed in time and space and can remain so for thousands of years. This permanence generated stable practices of discipline and attention in writing and reading, and these were the basis for the development of learning and knowledge production and consumption. Writing and reading promoted skills such as the capacity to identify themes and concepts and to analyze and draw inferences from them. The structure and grammar of words with their layers of meaning fostered human engagements with each other and in the world. And over historical time these engagements developed in and through education systems and the correlative swelling of mass literacy.

This specific interaction, neuroscience tells us, has had direct physical and neurochemical impact on the brain's architecture. Functional magnetic resonance imaging (fMRI) shows that reading printed text serves to strengthen the synaptic connections in the brain that deal with memory and concentration. And so not only does the analog connection to writing and reading enable us to become literate, discerning, and knowledgeable, but it also shapes the physical and neurochemical structure of our brain in ways that reflect this specific human-technology interactivity.

It's often forgotten that writing and reading have their own broad temporal pace and limits. We can work only so fast. Reading functions effectively only up to a certain point beyond which comprehension (recognition) begins to break down. This physical and cognitive fact is now becoming important. Digitally produced words on electronic screens are increasingly how we engage with writing and reading. Yet digital is a diametrically different technology that we nonetheless tend to treat as just a much-improved version of analog. However, in terms of cognition and brain functioning, reading LCD-generated words on an LED-backlit screen could hardly be more different from reading words on paper.

Eye-tracking software reveals how we engage with words on a screen. What researchers call the "F-shaped pattern" is how we tend to read screen content. As we read down a page, we are likely to read fewer and fewer words on each line, with the number gradually tapering off as the eye runs down the screen page to form the "F" pattern. There are several rather bleak reasons why we do this. First is that the browser itself is engineered specifically to keep us online and to keep our eyes moving while online. This is the business model that responds to the imperatives of the "attention economy." The more we stay online, moving from page to page, the more data is harvested by Google, Facebook, Amazon, and others, data they customize as user profiles for the advertisers on which their business depends. Our "attention" signifies an objective source of monetizable data for the platforms, but our interaction with computer screens signifies a subjective "distraction" that makes concentration and memorizing more difficult. And so, when reading online, as Mark C. Taylor tells it in the *Chronicle of Higher Education*, "Complexity gives way to simplicity, and depth of meaning is dissipated in surfaces over which fickle eyes surf. Fragmentary emails, flashy websites, tweets in 140 characters or less, unedited blogs filled with mistakes. Obscurity, ambiguity, and uncertainty, which are the lifeblood of art, literature, and philosophy, become decoding problems to be resolved by the reductive either/or of digital logic."[9] Recall the transformation of Nietzsche's writing toward "the telegram style" when he switched from pen to typewriter.

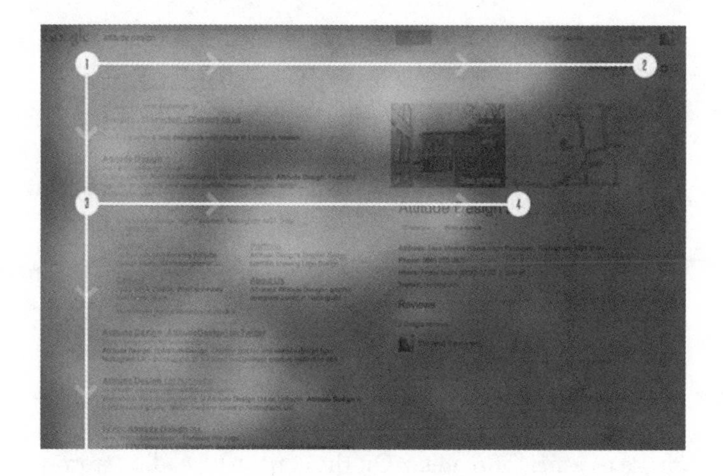

Figure 19 F-shape reading pattern of web content

When reading from screens, our eyes glide over web pages whose content and meaning attenuates in our heads due to our engagement with a screen engineered in part by algorithms to glide over web pages. Indeed, the technology engages more *with us* as opposed to us *with it*—as it manipulates, distracts, cajoles, and nudges us toward ever more content that we can never hope to keep up with. Growing disengagement with digital texts on a deliberately distractive interface means that we now experience what Herbert Simon, a forerunner in artificial intelligence and the psychology of human cognition, called a "poverty of attention." Distraction is driven, he argues, by the cognitive need "to allocate that attention efficiently among the overabundance of information sources that might consume it."[10] We struggle with too much information, in other words. Simon wrote this well before the Internet, before powerful web browsers, and before an engineered attention economy that competes for our eyes with information sources that are far greater in number and volume than Simon could ever have imagined.

When faced with this exponentially burgeoning volume of information, a common coping strategy is to skim-read and produce the "F" pattern that the eye-tracking software reveals in the lab. Skim-reading, however, doesn't make us efficient readers. On the contrary, reading from a screen has a visible effect on the synaptic architecture in the brain's hippocampus. Back in the lab, fMRI scans on the brains of predominantly digital readers show clearly that the synaptic connections relating to concentration and memory become weakened and atrophied as compared with print readers.[11]

Such findings could be taken positively, I suppose. We could interpret the redundancy of parts of the memory function of the brain as being simply the technology doing the work for us, relieving our brain in the same way that the automobile relieves our muscles. With computers and phones at our fingertips, we no longer need this capacity in our heads. Knowledge now increasingly resides in databases, within invisible servers at unknown locations. And knowledge becomes, like much else, digital, subject to automation, available on cybernetic feedback loops of call and response, like the GPS app on the mobile phone that tells us where we are through geolocation, and all we need do is touch the icon and look at it. No tacit knowledge or personal memory of where we are is required.

For the boosters of automaticity this is good news. The app on the mobile phone in place of memory acquired through doing means that the brain is relieved of superfluous knowledge of streets and directions and is freed up to do more creative things. Moreover, the argument goes, the moral panic about humans becoming superficial because of the corruptions of technology has been with us since (once again) the time of the Greeks, and from Plato in particular, who argued in dialogue with Socrates in the *Phaedrus* that the recent technology of writing was a curse and warned: "This discovery of yours will create forgetfulness in the learners' souls, because they will not use their memories; they will trust to the external characters and not remember themselves."[12]

Nicholas Carr updated this warning for the Internet age (and as a defense of print) in his book *The Shallows: What the Internet Is Doing to Our Brains*. In it he wrote about the forgetfulness-implanting effects of screen-based reading and argues, à la Plato, that "what the net seems to be doing is chipping away my capacity for concentration and contemplation. Whether I'm online or not, my mind now expects to take in information the way the net distributes it: in a swiftly moving stream of particles."[13]

Doomsday predictions, from Plato to Carr, about the harmfulness of new communication technologies have never come true. The sky remains up and we are still here in all our creative diversity and are still putting the vast amounts of knowledge we have acquired over the millennia to the service of the species itself: We've never lived for longer, never been healthier, never enjoyed so much material wealth; we've banished much disease and eased so much abject poverty and we've even stood upon the surface of the Moon, so what's the problem? they ask.

Well, if we accept the difference between analog and digital forms (of writing in this case), and appreciate, as McLuhan and many others tell us (and neuroscience confirms) that we shape our technologies and our technologies shape us in turn, then the sky will not indeed collapse onto our heads just because we read LCD texts from LED-backlit screens. But the fact that we do write and read more in this way means that we must ask some questions relating to the analog-to-digital transformation of our writing and reading practices. At a deeper philosophical level, screen-reading suggests a new relationship with knowledge, which is to say, a new interface in the constituting of the reality of the world. The fundamental question therefore is the status of knowledge—and how it is differently composed within each technological form. The way that the political process has changed through technological transformation from analog to digital is a useful case study to show what is at stake in this transformation.

Analog Democracy: Time for an Upgrade?

For tens of thousands of years human-technology interaction was elementary and developed slowly. Knowledge produced by this relationship was transmitted orally or passed on by seeing and doing within tribe or group. With the invention of writing, abstract and fixed knowledge was possible and could be shared through hand-copying and later machine-printed text. In Europe where printing and mass literacy emerged first, the technology of print and print culture had the power to legitimate ideas and disseminate knowledge oriented toward instruction and toward pathways to remedies for problems or challenges within society. Science, for example, became an important method of legitimation and carried within it the ideology of progress through technological solutions. Print generated and legitimated ideas in religion and philosophy and politics too. And through hundreds of years of disputation between defenders and attackers of each of the branches of knowledge, through wars of conquest and repression, through revolution and counter-revolution, sometime around the seventeenth century we arrived at a critical period in history: the Renaissance and its ancient theology gave way to the Enlightenment and its own civil religion. This was achieved in no small part through analog print culture. New politics converged with new ideas and were activated and propagated via print and literacy. A so-called republic of letters emerged to help give form and shape to a modern political philosophy. This was a wide-ranging, intense, and long-lasting correspondence between a critical mass of transatlantic political thinkers—an information network of thinkers that extended from Voltaire and Jean-Jacques Rousseau to John Locke and Benjamin Franklin, to name just some. Many more joined in the epistolary conversation, of course, through reading the thoughts of influential people, and then writing about them or acting on them. Journals, pamphlets, tracts, books, salons, and debating societies flourished to institutionalize and to a certain extent democratize these ideas through the growing literacy of industrializing societies. Such communication helped develop the central legitimating ideas of national-republican-liberal-democratic governance that are with us still today.

Modern politics is fundamentally based on the culture of print. And politics is at root an analog culture based on the analog extensions that science and technology, as evolving expressions of print, made possible—from pamphlet to newspaper, from telegraph to telephone and from movable type to mass media. It was on this point of the centrality of technology in the political process that the French philosopher Régis Debray laid stress on what he termed "the material forms and processes through which [political] ideas were transmitted—the communication networks that enable thought to have social existence."[14] He called this the "graphosphere," or the domain of writing. In the political field, and through the graphosphere, what we must now call "analog democracy" formed out of the Enlightenment and the "material forms and processes" that gave its ideas "social existence." In a sense, analog democracy is perhaps the most important human achievement of all; not that it is necessarily the moral and ethical pinnacle of all forms of human organization, but simply because it has been the most consequential organizing system that we have yet created.

Present-day democracy takes many forms, but they are based on the same communication technologies that enabled their distinctive expressions in the eighteenth century. Modern democracy began with the printed word whose ideas and material forms resonated with their creators and adherents through narratives that were recognizable to them in their physical communities. The abstract became material through their enaction in the local, regional, and national context of people, more or less in contiguity, through a relative closeness in time and space that has been called the "public sphere."

Figure 20 *The Reading Room*, Johann Peter Hasenclever, 1884

Analog material forms as expressed through print media, transportation, and communications, and based on the technologies of the time, shaped the methodology of politics. We recognize the method today, especially in formal political processes where laws are formed and enacted. What were these political processes? Based on the media of writing, reading, speaking, and listening, they were the interaction of people within the context of political institutions such as parliaments and congresses, with their committee rooms, speech-making daises, debating chambers, and other opportunities for interaction. This was *contiguous* interaction that produced a particular set of rhythms that correspond to the physical and cognitive capacities of the actors themselves and whose ideas, projects, and passions were given "social existence" through the technologies and techniques of the time. Grassroots politics was even more localized and contiguous, with the demonstration or political rally, elections and electioneering, campaigning and public speaking all playing an indispensable part not only of the functioning public sphere but also of the communal element of it that originated in the Greek *demos*. These are the dynamics of essentially analog processes that formed at a specific time in history through specific media that set down its specific temporal rhythm.

In the age of radio and television these traditional analog processes of institutional politics and public sphere participation endured. But the public sphere had changed. Its mode and expression had become increasingly analog-electronic. Analog newspapers still drove this process, but radio and television amplified the forms and processes and made the social existence of the reality of the world more diffuse and abstract. Nonetheless, even in the age of radio and television dominance, the local, national, and international news about the world and the political issues of the day were still measured in newspaper column inches. These contained stories about "what happened" in the world and the stories narrated the "facts" of the matter, communicated to us by a cadre of (usually) trusted journalists who were bound by a code of ethics—trained professionals whose primary skill, apart from journalistic enterprise and curiosity, was writing stories. Facts could always be disputed. However, reasoned debate, heated argument, the uncovering of countervailing facts, or the exposure of lying or a cover-up could settle the matter. That facts existed somewhere "out there" was not in dispute because our whole print culture tradition of science and Enlightenment reason told us they did. As public-sphere media, radio and television stretched the tenability of the process because the recognition and contiguity factors were no longer so clear and close. However, analog-electronic mass media continued as the dominant form right up until the beginning of the twenty-first century.

We can no longer easily participate in the political process as we did even a generation ago. Digital forms and processes are so different that we do not easily recognize them, yet we blithely trust in them to work for us. Mainstream media (what's left of it) still tells stories about politics. It still writes the narratives in which the world should make sense to readers and writers through the facts that the stories contain. But digital communication functions differently from analog. Platform-owned algorithms are formed on a specific business model, whose code is a closely guarded secret. These algorithms filter, profile, select, and distribute these stories, these facts, in ways that have served to distort the political process we call liberal-democratic. This is no exaggeration. The details of this claim are for another, more specialized book. Here I will give only a brief insight into what is perhaps the greatest challenge of the digital age, which is that facts and knowledge, and the politics that depend on these for their legitimacy, are all wrapped up in a new and negating communication paradigm.

Since around 2016, with the election campaigns of Donald Trump in the United States and the referendum on Brexit in the United Kingdom, the effects of private algorithms operating through the public sphere have become clearer. Revelations that social media can transform the very core of what constitutes reality in the heads of hundreds of millions of people across the planet came as a shock to legions of users who treated the "free stuff" of social media with naive complacency. For much of its content, social media still relies on mainstream media. However, that these media have moved online serves mainly to make their connection to the wider digital sphere more seamless. "Mainstream" is therefore a misnomer, as it sits within a digital spectrum that contains every shade of fact, knowledge, and opinion that is mixed up with untruth, half-truth, and invention.

Within the different jurisdictions in which they operate, powerful platforms like Facebook, Google, and Twitter lobby constantly to ensure that they are not classified as publishers, as are *Le Monde*, Melbourne's *Herald Sun*, or the *Baltimore Post-Examiner*, for example. This means that they are not legally or editorially responsible for the stories they disseminate through their networks. As just noted, we now know, and mainly because of old-fashioned investigative journalism, something about how these algorithms work. They organize data through code designed to attract users online and manipulate them to keep them online. Using psychology and powerful algorithms to profile users so that it "knows" what is relevant and meaningful to them, data companies can shape the consciousness of the individual to an unprecedented degree. In the political sphere, they know, for instance, that people are attracted to controversy in media, and so social media algorithms search for what presses your political buttons and will suggest more and more of it for you. Likewise, your profiles show your political interests and biases, and these too are continually reinforced. The result is the "filter-bubble" effect where users "find" like-minded souls to share like-minded content with.

The digital public sphere is thus a constellation of largely private spheres that rarely trespass upon each other's domain. To be heard within these spheres, to be a so-called thought leader within them, means having to ratchet up levels of controversy and noise in their contributions and their sharing, contributions that can easily slip over into more shrill and dogmatic beliefs about the world. Contrary or alternate views are in danger of being seen to be beyond the pale of decency and reason, and so trolling or doxing or textually abusing not only is an easy response to views that are different, but serves to harden the borders between publics still further.

Social media networks are manipulable for those with the requisite computer skills and political organization. The wider sphere becomes vulnerable to misinformation proliferated by "bad actors" such as hostile states and subversive or extremist political groups or even individuals. Misinformation and distortion flow through the same networks as fact-checked and transparent sources, to create rivers of polluted information that makes politics an unreal process where many are skeptical about much of what they see. Cheap fakes and deep fakes multiply, textually as well as visually, to the point where off-the-shelf machine-learning software can synthesize voices and faces, and where lip-synching and body gestures can be manipulated to say and do almost anything. This uncertainty can push people deeper into their own domains, toward their own identity groups, and to believe in them even more strongly.

The digisphere cannot be a public sphere of the newspaper and coffeehouse kind that was envisaged as the founding communicative form of modern democracy. The human association with writing, reading, and words has changed utterly. And for the first time since the invention of writing and the circulation of literacy, what constitutes

Figure 21 Cover of *The Atlantic* magazine, June 2020

knowledge and fact, the very things that writing first established, is no longer so clear, to anyone.

Is Analog Democracy Upgradable to Digital?

In theory, as we've seen, automation is set up to relieve the burden of labor. In practice, and across an ever-widening social scale, it serves to separate us from having an active role in the building of our material culture. And as noted also, this has the effect of making us sedentary and therefore liable to diabetes, cancers, and other lifestyle diseases. Automation also disconnects us from our ancient and deeply rooted connections with the physical world and the natural environment. Looking at screens through which to live one's life, toward an existence like those depicted in *WALL-E*, cannot be said to constitute progress. The film's inner message is that technology has gotten away from us. We trashed the planet and are now lost forever in space, sustained by but also dominated by machines that were meant to sustain and maintain both humans and environment.

More importantly, perhaps, is the cognitive transformation, the digital consciousness derived from pursuing the political process online. Democracy of the analog sort is impossible through an increasingly toxic digital social media environment. This poses a dilemma. As a society we must decide what kind of politics is possible to effect positive change in the world.

If it is to be through the digital sphere, then individual and collective control over it is necessary. That would require public control, or public accountability, over how the facts about the world are generated and disseminated, a publicly controlled or socially transparent algorithm whose logic and ends are clear: a public social media with a public charter.

If we decide that the analog political process we inherited from the Enlightenment should be preserved and should lead us into an increasingly digital twenty-first century, then a different kind of digital mass media is

Democracy of the analog sort is impossible through an increasingly toxic digital social media environment.

going to be necessary to accommodate this. If we want to see and acknowledge truths and facts about the reality of the world, then this is going to take a form of social media with responsibilities not just toward profits and shareholders but also toward the public sphere. That would mean an enlarged and enhanced role for publishers, with gatekeepers who are editors, trained and professional, and with an analog ethics of responsibility toward truth and how it's reported and debated.

Neither of these scenarios seems likely at the time of writing. So, we need to ask ourselves a few basic philosophical questions about our relationship to technology in a phase of history where technological change has never been so consequential, and where two separate categories of technology permeate our world.

Analog needs to be understood more widely. This means seeking to understand ourselves better, so that the mysteries of the seductiveness and authenticity of disappearing analog technologies will be seen for what they are, ancient aspects of who and what we are. Our technological roots and heritage must be a treasure that we value and nurture and not regard as obsolete and inefficient.

Digital we hardly understand at all, but mostly we unthinkingly act as if it is the solution to all our problems.

Whatever the future holds, we will continue to communicate through the telling of stories as we have always done. Because that is how we evolved. However, reflecting

Figure 22 (Left) The *Guardian* newsroom, London, 2014 (a blend of analog and digital); (right) Facebook server farm, inside Sweden's Arctic Circle, 2017 (wholly digital/automated)

the dominance of digital communication today, there is the danger that our stories will be shorter, angrier, and more demanding. And driven by social media, there is the risk that they will gravitate further toward tribalism and identity politics, toward hostility and suspiciousness, toward paranoia and intrigue. And these stories will be stories sure in their own rightness and truth—and certain in the wrongness and falsity of the stories of others not recognized as part of the group. The underlying existence of truth and facts is becoming a concept not universally shared. Truths can have counter-truths, and facts met with alternative facts.

The body and brain become increasingly susceptible to such an increasingly strange and unrecognizable world. But our limitations and weakness in the face of new and powerful technology are not new. In his essay "The

Storyteller," from 1936, Walter Benjamin told of the dwindling of the skills of oral and written storytelling, of the decline of the ancient retelling of human experience through intricately remembered and finely crafted narratives. We moderns have been impoverished by technological "progress," he argued, much of it derived from the science of warfare. And we should not forget that our Internet is a spin-off from Cold War military research. The degeneration is also a spiritual one, Benjamin adds, a decline that stands no chance when set against material power that has developed beyond actual human interests. In his essay Benjamin reflects on our vulnerability to the new and increasingly autonomous analog technologies such as machine guns, tanks, and airplanes, technologies that caused such unprecedented carnage in the Great War: "A generation that had gone to school on a horse-drawn streetcar now stood under the open sky in a countryside in which nothing remained unchanged but the clouds, and beneath these clouds, in a field of force of destructive torrents and explosions, was the tiny, fragile human body."[15]

That fragile human body was once resonant and in natural harmony with technology. For millennia the modest extensions through which we survived left little trace on, nor made any great demands from, the environment of which we have always been a part. Analogicity, resonance, equivalence, and recognition have been our expressions with the world for most of our time on Earth. But we slowly, then rapidly, evolved with the things we made. These made us clever, and we transferred this cleverness back into the next, even cleverer, invention. Increasingly complex technologies became increasingly distant from us, stretching our recognition of what they do almost to breaking point around the last quarter of the twentieth century when analog-electronic mass media began to make everyone "present and accessible" to everyone else at the same time, as the great Marshal McLuhan told us. The "black box" of digital magic tricks stood ready to take over.

We were entering a new phase of non-evolution with a category of technology that is not us. Where this takes us is increasingly out of our hands, so autonomous is our digital creation becoming. But we still have the potential of politics and of will and reflection and intelligence. If we can harness these toward the human essence of analogicity and ends of interaction with technology—even digital technology—then the planet may return to something like a balance. There's never been a more urgent time to seek that balance, that connection and resonance with technology, with society, and with, above it all, the environment. And it is from that point of holistic essential knowledge about analog and digital that we could begin to shape our world to become a more sustainable place, a place not only for us but also for the flora and fauna and air and weather systems that since modern times we've viewed merely as a resource to feed our awesome yet fated technological prowess.

GLOSSARY

Alienation

Linked to the experience of modernity, alienation signifies a psychological or social malady, one involving a negative estrangement between a self and other that properly belong together.

Analog

Analog is popularly thought of as machines, devices, and technologies that are predigital. However, the understanding here is that it is much more. The *OED* defines analog as "a thing which (or occasionally person, who) is analogous to another; a parallel, an equivalent." This more encompassing definition is extended here to define the predigital relationship between humans and technology and that humans were themselves embodiments of the technologies they created, from simple hand tools to complex electronic machines.

Analog as "recognizable" technology

Following from the previous definition, and drawing from philosophical anthropology, a feature of analog technology is that it is "recognizable" in that it constitutes an activity that functions in a visible way that allows us to grasp the link between a movement and its effect, the process, and the continuity.

Analog embodiment

Analog embodiment is a concept that states that humans are themselves technological creatures who survived because they evolved with tool-use and toll-inventing capacities. Humans are embodied (analog) technology.

Authenticity

A commonly attributed quality to analog tools and processes, especially in the "retro" analog cultural markets, authenticity here means the feeling that some analog technologies are more "real" and closer to the source; for example, a vinyl record is more "real" and closer to the source than is a digital MIDI file that is the basis of a compact disc recording.

Automation
Automation signifies the production of goods, services, and processes with as little human input or participation as possible. In the current age of digital domination of production, it signals a more profound shift away from the human-technology relationship of co-creation, and toward a disconnect of alienation from the digital technology and the goods, services, and processes it creates.

Digital
Digital refers to machine processes and forms of logic that are based on the principle of discrete or discontinuous data values; it is the primary logic of computation. It is contextualized here as the opposite of analog processes and forms that are based on principles of "continuity" and "flow."

Digital as a form of "magic"
Digital as a form of "magic" relates here to the definitions of digital and the affordances of digital that work at scales and speeds that are impossible to register humanly, and impossible for ostensibly similar analog processes to compete with. The meaning here is drawn from the influential futurist and inventor Arthur C. Clarke, who wrote in 1962, "Any sufficiently advanced technology is indistinguishable from magic."

Electronics
Electronics are distinguished from "electricity" and its current flow. Electronics enables the controlling of current flow. Electronics gives the capacity to "switch" the flow of current to make it act in more complex ways and with more sophisticated effects. The meaning here emphasizes the fact that electricity powers machines, whereas electronics enables them to *make decisions*. These properties paved the way for computational logic to be driven by electrical power.

Enlightenment
An intellectual and political movement that began in Europe and North America in the late seventeenth century, coming to a flowering a century later, Enlightenment was characterized by a commitment to, and emphasis on, the primacy of science and reason as opposed to religion and monarchical rule. It provided the intellectual and cultural basis for the rise of capitalism and modernity.

Graphical user interface (GUI)
GUI is a digital interface that allows easy access, navigation, and manipulation of computer operating systems. Devised in the late 1960s by the Xerox Parc company, GUI applications were taken up and popularized by Apple and Microsoft, among others, to become the technical basis with which most people interfaced with computers. Typical examples today are the app icons on a mobile phone.

Human-technology relationship
Encompassed by a broad academic research program that through anthropology, philosophy, science and technology studies, and media archaeology, the human-technology relationship speculates, among other concepts, that humans have always been technological creatures and, further, that the analog tools and processes they invented are embodiments of themselves and vice versa.

Jacquard loom
The Jacquard loom was a major advance in mechanical cotton weaving in France in the early nineteenth century. Its major innovation—based on earlier inventions—was a series of wooden cards strung together, each having holes punched into it that would correspond to a row in the design of the textile being weaved. The card system was in effect a "code" that would direct the pattern of the textile and allowed a uniform outcome of increasingly complex design. It acted as a significant de-skilling innovation for manual weavers. It also became the basis of early machine computer designs by Charles Babbage later in the century.

Malling-Hansen writing ball
A mid-nineteenth-century typewriter and early example of innovation in "mechanical writing," the Malling-Hansen writing ball consists of fifty-two keys attached to a hemispherical ball that punched the letters onto a cylinder where the sheet of paper would be held. One of the most famous users was the near-blind Friedrich Nietzsche, who bought one in 1882 and wrote all his subsequent books on this and other typewriters.

Mellotron
The Mellotron is an electronic-analog keyboard-based musical instrument developed in 1963 in England. When a key is pressed, it pushes a magnetic tape with the recorded sound (of various instruments) around a capstan and over a magnetic tape head to produce the sound.

Modernity
Modernity describes a historical period that began with the Industrial Revolution and the Enlightenment. It encompasses many developments in human thought, culture, economy, and society that arose in the eighteenth century, to become more fully formed through the twentieth. Features such as rationalism, secularism, individualism, and urbanization all served to shape a modern consciousness that saw such concepts as progress, rapid change, material wealth, and technological innovation as embodiments of the modern spirit.

Natural user interface (NUI)
A technical advance on the graphical user interface (GUI) of earlier versions of personal and networked computers, NUI works on "intuitive" use, where interaction comes easily and with the minimum of delays or "friction." Examples we see today in tablet computers or voice- activated "assistants" such as Siri or Alexa need virtually no instruction but are set up in such a way that even very young children can explore and exploit the device readily.

Philosophical anthropology
Dealing with questions of human behavior as reflective of the social, natural, and technological environments, philosophical anthropology is underpinned by a phenomenological approach that uses the structures of experience, consciousness, and imagination to understand the situatedness of the individual in relation to social, environmental, economic, and technological dynamics.

Technological determinism
Technological determinism is the idea that the technologies we use determine the broad shape of society, economy, culture, class structures, and the form and function of our material world. Marshall McLuhan's dictum "The medium is the message" is the classic formulation of the concept, whereby it is not the "content" of the medium that is so important but what the technology itself enables and determines. An example might be the automobile, which has been claimed since its invention to be a technology of individuality and freedom, but, in fact, mostly acts as a determining force for the shape of cities, industries, and the mass regulation of human movement.

Technology

Technology denotes the material and nonmaterial "extensions" that humans have developed as a key element of their survival as a species, from the simple hand tools of prehistoric times to the complex machines and processes, both analog and digital, that produce the goods and services that shape our world today. The idea that humans are themselves technology and that it is this quality and capacity that differentiate them from other species is a primary contention of this book.

Telegraph

The word "telegraph" is a combination of the Greek *tele*, meaning "distant," and *graphein*, meaning "to write." The invention of the telegraph in the mid-nineteenth century was functionally an improvement of the semaphore system of communication, versions of which had been in existence as aids to naval communication since the late eighteenth century. The telegraph is an electrically powered coded signal that transmits down a wire, or through radio waves, to be decoded at the other end. It revolutionized human communication across many areas of activity and, according to Karl Marx, effected the "annihilation of space and time" and provided an important technological basis for the first phase of capitalist imperialism and globalization over the late nineteenth century.

NOTES

Chapter 1

1. Bill McKibben, "Pause! We Can Go Back!," review of David Sax's *Revenge of the Analog*, *New York Review of Books*, February 9, 2017.
2. William Gibson, *Mona Lisa Overdrive* (London: Orion), 239.
3. Killian Fox, "What Does a £2,500 Record Sound Like?," *Guardian*, accessed October 19, 2021, https://www.theguardian.com/music/2013/may/25/pete-hutchison-interview-new-vinyl-recording.

Chapter 2

1. Chris Cheshire, "The Ontology of Digital Domains," in *Virtual Politics: Community and Identity in Cyberspace*, ed. David Holmes (London: Sage, 1997), 40.
2. Jonathan Sterne, "Analog," in *Digital Keywords: A Vocabulary of Information Society and Culture*, ed. Benjamin Peters (Princeton NJ: Princeton University Press, 2016), 32.
3. OED, "Analog," accessed October 25, 2021, https://www.oed.com/view dictionaryentry/Entry/7029.
4. OED, "Analog."
5. Samuel Johnson's *Dictionary of the English Language*, 6th ed., vol. 1 (London: J. F. and C. Rivington, 1785), 85, accessed September 15, 2021, https://publicdomainreview.org/collection/samuel-johnson-s-dictionary-of-the-english-language-1785.
6. Johnson, *Dictionary*, 1132.
7. Arnold Gehlen, *Man in the Age of Technology* (New York: Columbia University Press, 1980), 4.
8. Gehlen, *Man*, 19.
9. Gehlen, *Man*, 14.
10. Silvia Estévez, "Is Nostalgia Becoming Digital?," *Social Identities*, 15, no. 3 (2009): 401.
11. Estévez, "Nostalgia," 402.
12. Estévez, "Nostalgia," 401.
13. R. W. Gerard, "Some of the Problems Concerning Digital Notions in the Central Nervous System," in *Cybernetics: The Macy Conferences 1946-1953. The Complete Transactions*, rev. ed., ed. Claus Pias (Chicago: Diaphanes Publishers/University of Chicago Press, 2016), 172.

14. Marshall McLuhan, *Understanding Media: The Extensions of Man* (New York: McGraw-Hill, 1964), 55.
15. McLuhan, *Understanding Media*, 7.
16. John Culkin, "A Schoolman's Guide to Marshall McLuhan" (1967), accessed September 9, 2020, https://mcluhangalaxy.wordpress.com/2017/09/19/a-schoolmans-guide-to-marshall-mcluhan-by-john-culkin-s-j-1967/.

Chapter 3

1. Nicholas Negroponte, *Being Digital* (New York: Hodder & Stoughton, 1995), 39.
2. Daniel Bell, *The Coming of Post-Industrial Society* (London: Heinemann Educational Books, 1974).
3. Crosley Radio website, accessed October 20, 2021, https://www.crosleyradio.eu/pages/about.
4. Alex Petridis, "The Crosley Generation: The Record Player That Has the Kids in a Spin," *Guardian*, April 21, 2016, accessed October 20, 2021. https://www.theguardian.com/music/2016/apr/21/crosley-generation-record-player-has-the-kids-spinning.
5. David Sax, *Revenge of the Analog: Real Things and Why They Matter* (New York: PublicAffairs, 2016).
6. Sax, *Revenge*, xvii.
7. Janna Anderson and Lee Rainie, "The Positives of Digital Life," Pew Research, July 3, 2018, accessed September 14, 2021, https://www.pewresearch.org/internet/2018/07/03/the-positives-of-digital-life/.
8. Richard Sennett, *The Craftsman* (London: Penguin).
9. Guy Raz, National Public Radio Interview with Matthew Crawford, July 12, 2009, accessed January 4, 2020, https://www.npr.org/transcripts/106513632.
10. Sennett, *Craftsman*, 20.
11. Carol Wilder, "Being Analog," in *The Postmodern Presence*, ed. Arthur Berger (London: Sage, 1998), 252.
12. Wilder, "Being Analog," 241.
13. George Lakoff and Mark Johnson, *Metaphors We Live By* (Chicago: University of Chicago Press, 1980), 3.
14. Richard Dawkins, *The Selfish Gene* (Oxford: Oxford University Press, 1989), 192.
15. Steven Levy, *Insanely Great: The Life and Times of Macintosh, the Computer That Changed Everything* (New York: Viking, 1992), 61.

Chapter 4

1. Helen Epstein, "The Highest Suicide Rate in the World," *New York Review of Books*, October 10, 2019, accessed July 3, 2020, https://www.nybooks.com/articles/2019/10/10/inuit-highest-suicide-rate/.
2. Yuval Noah Harari, *Sapiens: A Brief History of Humankind* (New York: Vintage Publishing, Harvill Secker, 2014), 137.
3. Walter Ong, *Orality and Literacy: The Technologizing of the Word* (London: Routledge, 1982), 73.
4. Ong, *Orality*, 81.
5. Ong, *Orality*, 80.
6. Walter Ong, "Writing Is a Technology That Restructures Thought," in *The Written Word*, ed. Gerd Baumann (Oxford: Oxford University Press, 1986), 23.
7. T. Freeth et al., "Decoding the Ancient Greek Astronomical Calculator Known as the Antikythera Mechanism," *Nature* 444 (2006): 587–591.
8. G. J. Whitrow, *What Is Time?* (London: Thames and Hudson, 1989), 128–129.
9. E. P. Thompson, "Time, Work Discipline and Industrial Capitalism," *Past & Present*, no. 38 (December 1967): 56–97.
10. Arran Gare, *Nihilism Inc.: Environmental Destruction and the Metaphysics of Sustainability* (Sydney: Eco-Logical Press, 1996), 104.

Chapter 5

1. Cited in Alvin Snider, "Cartesian Bodies," *Modern Philology* 98, no. 2, Religion, Gender, and the Writing of Women: Historicist Essays in Honor of Janel Mueller (November 2000): 309.
2. Madeline M. Muntersbjorn, "Francis Bacon's Philosophy of Science: Machina Intellectus and Forma Indita," *Philosophy of Science* 70, no. 5 (2003): 1145.
3. E .T. Bell, *Men of Mathematics* (New York: Simon and Schuster, 2014), 123.
4. E. P. Thompson, *The Making of the English Working Class* (New York: Vintage Books, 1966), 279, 312.
5. Aristotle, cited in Roger Boesch, "Aristotle's 'Science' of Tyranny," *History of Political Thought* 14, no. 1 (Spring 1993): 8.
6. Karl Marx, *The Grundrisse* (New York: Vintage, 1973), 705.
7. Rahel Jaeggi, *Alienation* (New York: Columbia University Press), 1.
8. Maryanne Wolf, *Proust and the Squid: The Story and Science of the Reading Brain* (New York: Harper Perennial, 2008).
9. Friedrich Kittler, *Gramophone, Film, Typewriter* (Stanford, CA: Stanford University Press, 1999), 205–206.
10. Kittler, *Gramophone*, 203.

11. Kittler, *Gramophone*, 203.
12. Kittler, *Gramophone*, 210.

Chapter 6

1. Available to view online at https://www.biodiversitylibrary.org/item/28001#page/557/mode/1up.
2. *Meno* by Plato, online at MIT Classics, accessed October 21, 2021, http://classics.mit.edu/Plato/meno.html.
3. "Letter from Benjamin Franklin to Peter Collison dated May 25, 1747," Benjamin Franklin Historical Society, accessed June 26, 2021, http://www.benjamin-franklin-history.org/letter-from-benjamin-franklin-to-peter-collison-dated-may-25-1747/.
4. Rebecca Northfield, "Francis Ronalds the Forgotten Father of the Electric Telegraph," *E&T Engineering and Technology*, July 12, 2016, accessed January 19, 2020, https://eandt.theiet.org/content/articles/2016/07/francis-ronalds-the-forgotten-father-of-the-electric-telegraph/.
5. Northfield, "Francis Ronalds."
6. See "Adverts for the Introduction of the Telegraph on the Great Western Railway in England," accessed February 11, 2021, https://www.alamy.com/adverts-for-the-introduction-of-the-telegraph-on-the-great-western-railway-in-england-image210380066.html.
7. A. D. Coleridge, *Goethe's Letters to Zelter* (London: George Bell & Sons, 1892), 246.
8. Karl Marx and Friedrich Engels, "The Manifesto of the Communist Party," in *Selected Works* (Moscow: Progress Press, 1976), 40.
9. Jacques Ellul, *The Technological Society* (New York: Vintage Books, 1964), 248.
10. Richard Sennett, *The Craftsman* (London: Penguin), 85.
11. R. W. Gerard, "Some of the Problems Concerning Digital Notions in the Central Nervous System," in *Cybernetics: The Macy Conferences 1946-1953. The Complete Transactions*, rev. ed., ed. Claus Pias (Chicago: Diaphanes Publishers/University of Chicago Press, 2016), 172.
12. J. C. R. Licklider, "Man-Computer Symbiosis," *IRE Transactions on Human Factors in Electronics* 1 (March 1960): 4–11.
13. Marshall McLuhan, *Understanding Media* (Corte Madera, CA: Gingko Press, 2003), 333.
14. McLuhan, *Understanding Media*, 86.
15. Michel Foucault, *Discipline and Punish* (New York: Vintage Books, 1977), 135–169.
16. Raymond Williams, *Television: Technology and Cultural Form* (London: Routledge, 1990), 98.

17. Williams, *Television*, 79.
18. Guy Debord, *The Society of the Spectacle* (New York: Zone Books, 1995), 12.
19. Debord, *Spectacle*, 11.
20. *Whole Earth Catalog* (Menlo Park, CA: Portola Institute, 1968), 2.
21. *Whole Earth Catalog*, 34.
22. Steven Levy, *Hackers: Heroes of the Computer Revolution* (New York: Doubleday, 1984), 165.

Chapter 7

1. "The Notebooks of Samuel Butler," accessed October 23, 2021, https://www.gutenberg.org/files/6173/6173-h/6173-h.htm.
2. Christof Koch, *The Feeling of Life Itself* (Cambridge, MA: MIT Press), 129.
3. Bernard Stiegler, "Technologies of Memory and Imagination," *parrhesia* 29 (2018): 75n47.
4. Robert MacBride, *The Automated State* (Philadelphia: Chilton Book Company, 1967), 4.
5. Sigmund Freud, *Civilization and Its Discontents* (New York: W. W. Norton, 1989), 44.
6. Arnold Gehlen, *Man in the Age of Technology* (New York: Columbia University Press, 1980), 14.
7. Brett Frischmann and Evan Selinger, *Re-engineering Humanity* (Cambridge: Cambridge University Press, 2018), 13.
8. Bryan Magee, *Ultimate Questions* (Princeton, NJ: Princeton University Press, 2016), 72.
9. Mark C. Taylor, "Speed Kills: Fast Is Never Fast Enough," *The Chronicle of Higher Education*, October 20 2014, accessed February 9, 2021, https://www.chronicle.com/article/speed-kills/.
10. Herbert Simon, "Designing Organizations for an Information-Rich World," in *Computers, Communications, and the Public Interest*, ed. Martin Greenberger (Baltimore, MD: Johns Hopkins University Press, 1971), 40–41.
11. Nicholas Carr, *The Shallows: What the Internet Is Doing to Our Brains* (New York: W. W. Norton, 2010), 23–25.
12. Plato, *Phaedrus*, accessed October 22, 2021, http://classics.mit.edu/Plato/phaedrus.html.
13. Carr, *The Shallows*, 10.
14. Régis Debray, "Socialism: A Life-Cycle," *New Left Review* 46 (July–August 2007): 5.
15. Walter Benjamin, "The Storyteller," in *Illuminations*, trans. Harry Zohn (New York: Schocken, 1968), 84.

FURTHER READING

Books on Analog and Analog Culture

Alben, Alex. *Analog Days: How Technology Rewrote Our Future*. Scotts Valley, CA: CreateSpace Independent Publishing Platform, 2012.

Burns, Jehnie I. *Mixtape Nostalgia: Culture, Memory, and Representation*. Lanham, MD: Rowman & Littlefield, 2021.

Horton, Zac. *The Cosmic Zoom: Scale Knowledge and Mediation*. Chicago: University of Chicago Press, 2021. See chapter 2, especially.

Krukowski, Damon. *The New Analog: Listening and Reconnecting in a Digital World*. New York: New Press, 2017.

Pinch, Trevor, and Frank Trocco. *Analog Days: The Invention and Impact of the Moog Synthesizer*. Cambridge, MA: Harvard University Press, 2004.

Reynolds, Simon. *Retromania: Pop Culture's Addiction to Its Own Past*. New York: Farrar, Straus and Giroux, 2011.

Sax, David. *Revenge of the Analog: Real Things and Why They Matter*. New York: Public Affairs, 2016.

Sterne, Jonathan. "Analog." In *Digital Keywords: A Vocabulary of Information Society and Culture*, ed. Benjamin Peters, 31–45. Princeton, NJ: Princeton University Press, 2016.

Wilder, Carol. "Being Analog." In *The Postmodern Presence*, ed. Arthur Berger, 239–253. London: Sage, 1998.

Books on Analog Technology

Howard, K. Shannon. *Unplugging Popular Culture: Reconsidering Materiality, Analog Technology, and the Digital Native*. New York: Routledge, 2020.

Murphy, Sheila C. "Shake It like a Polaroid Picture: The Rise and Fall of an Analog Social Medium." In *The Routledge Companion to Media Technology and Obsolescence*, ed. Mark J. P. Wolf, 234–242. New York: Routledge, 2019.

Rocchi, Paolo. *Logic of Analog and Digital Machines*. Hauppauge, NY: Nova Science Publishers, 2011.

Small, James S. *The Analogue Alternative: The Electronic Analogue Computer in Britain and the USA, 1930–1975*. New York: Routledge, 2002.

Weyrick, Robert C. *Fundamentals of Analog Computers*. Hoboken, NJ: Prentice Hall, 1969.

Papers and Articles

Bartmanski, D., and Woodward, I. "The Vinyl: The Analogue Medium in the Age of Digital Reproduction." *Journal of Consumer Culture* 15, no. 1 (2015): 3–27.

Cauduro, E. "Photo Filter Apps: Understanding Analogue Nostalgia in the New Media Ecology." RIMAD Research Institute for Media, Art and Design. University of Bedford, 2014. Accessed November 11, 2021. https://uobrep.openrepository.com/handle/10547/603550.

Epstein, Dmitry, Mary Newhart, and Rebecca Vernon. "Not by Technology Alone: The 'Analog' Aspects of Online Public Engagement in Policymaking." *Government Information Quarterly* 31, no. 2 (2014): 337–344.

Haugeland, John. "Analog and Analog." *Philosophical Topics* 12, no. 1 (1981): 213–226.

Holmes, N. "Digital Machinery and Analog Brains." *Computer* 44, no. 10 (October 2011): 100–199.

O'Hagan, S. "Analogue Artists Defying the Digital Age." *Guardian*, April 24, 2011. Accessed November 10, 2021. https://www.theguardian.com/culture/2011/apr/24/mavericks-defying-digital-age.

Sacasas, L. M. "The Analog City and the Digital City." *New Atlantis*, no. 61, Center for the Study of Technology and Society (2020): 3–18.

Sterling, B. "The Analog Culture Destroyed by Digital Culture Does Not Come Back." *Wired*. Accessed November14, 2021. https://www.wired.com/2012/11/the-analog-culture-destroyed-by-digital-culture-does-not-come-back/.

INDEX

Page numbers followed by f indicate figures.

ROBERT HASSAN is Professor of Media and Communications at the University of Melbourne. He is the author of eleven books, including *Philosophy of Media* and *Uncontained: Digital Detox and the Experience of Time*.